# FIERCE ENCOUNTER

*Life and death in the Australian bush*

*by* H. D. WILLIAMSON
*illustrated by* WILLIAM COOPER

A. H. & A. W. REED • Sydney • Melbourne • Wellington • Auckland

# Fierce Encounter

*Life and death
in the Australian bush*

*First published 1970*

A.H. & A.W. REED

51 Whiting Street, Artarmon, Sydney
357 Little Collins Street, Melbourne
182 Wakefield Street, Wellington
29 Dacre Street, Auckland

ISBN 0 589 07085 1

*Filmset in 11/13pt Apollo by Craftsmen Type-Setters Pty Ltd, Sydney*
*Printed and bound by Dai Nippon Printing Co (International) Ltd, Hong Kong*

# Contents

# Ungalla the Hunter

IT was still dark when Ungalla awoke and rolled over to stare drowsily at the red points in the little heap of embers at his side. He picked a few twigs from a nearby pile, and as he dropped them on to the coals a flame broke through, glinting on his smooth black body and broad-nosed, thick-lipped face.

He sat up. A half-grown dingo stretched lazily, then sidled up and nosed at him. Most of the other fires in the camp were cold, but, here and there a thin wisp of smoke spiralled into the night sky. A piccaninny cried and was cuffed again into silence by a sleepy lubra. The wind that often precedes the dawn stirred the smoke curls as Ungalla collected his spears and woomera and rose to his feet, a film of powdery ash floating down from his shoulders.

Moving carefully between the huddled forms of lubras, children, warriors and bearded ancients, he bent to touch Milunga lightly on the arm. Thus summoned, the younger warrior was swiftly alert. He seized his spears and followed Ungalla into the bush. Bright as the stars were overhead, the light that reached the ground seemed no more than a lesser degree of darkness. Yet it was enough, and the Aborigines moved with a steady swiftness, a pace that was unhurried but unvarying—though occasionally Ungalla or his companion would pause to gather a few berries from a bush. So they breakfasted, meagrely it is true, but with murmurs of satisfaction and the smacking lip of relish.

They turned along the base of a ridge. Grass trees with single flower spikes pointing upward stood in silhouette against the pale sky like warriors, armed and watchful.

The two men stole deftly on. There was a greyness about the eastern horizon, and faint colour tinged a lonely cloud.

Ungalla pointed to the ground and swung away at right angles. The trail was fresh, and he followed it with the restrained eagerness of the experienced hunter. Here a tussock was cropped short where the quarry had paused; a little farther on a broken twig, a paw mark, told its story. In broad daylight a white man would have doubted many of the signs even after they had been pointed out to him, but the Aborigines read them unerringly.

They had travelled two or three miles before Ungalla motioned to his companion to stay and climbed a low ridge to examine the shadowed valley on its far side.

Under the lee of a great boulder he waited, still as the stones around him.

Drops of dew glittered on a nearby shrub. Vague blurs took shape —boulders, bushes, spinifex. And at the far side of the hollow a dark shape hopped to a fresh patch of feed.

The watcher counted three kangaroos—a buck, a doe and a joey. He backed slowly down the ridge, beckoned to Milunga, and retraced his steps to a shallow gutter cut by the rains. The dry watercourse zigzagged across a gap in the ridges. The two dropped down and, crouching low, sped to the shelter of the next rise.

Ungalla slowed and stopped.

Milunga grunted understandingly and remained as the other fitted a spear to his woomera.

When he had made sure that the hole in the butt of the spear was gripped by the tooth in the end of the throwing stick and that the shaft lay snugly in its concave length, he tested his grip again and steadied the front of the spear with his left hand.

Then, already in the throwing position, he drifted up the slope like a shadow. His head rose almost imperceptibly above the ridge.

The kangaroos were about two hundred yards distant, the buck slightly to the side of the others.

A light wind brought their scent to Ungalla's nostrils. His right arm swayed slowly back and he stood like an ebony statue, the woomera— that extra and more powerful joint in the hunter's arm—lying beneath the spear shaft as the shin of a squatting warrior folds beneath his thigh.

A kangaroo sat up on his great haunches and looked through the confusing semi-tones of grey. As he again lowered his head to feed, Ungalla ran noiselessly closer, keeping a boulder between himself and his goal. When the quarry again glanced up, the hunter was invisible. A second soundless rush took him halfway across the remaining distance. He stopped, his arms still upheld, his spear poised, his body set.

8

With uncanny anticipation he was always stationary when a kangaroo raised its head from the grass. His advance became less rapid as he closed in. There was a catlike stealth in every cautious step, and the fatal arm seemed locked in its menacing pose. If the position was uncomfortable, he showed no sign of distress.

Concealment became more difficult. Now even the smallest sapling was better than no cover at all, and, at last, he was standing within the same open space as his prey.

Again the buck kangaroo gazed inquiringly round, and this time he did not return to the pleasurable business of eating as readily as before. His big ears moved alertly. He tested the breeze as though some sixth sense warned him of a danger he could not see or hear or scent. The doe and the joey fidgeted, infected by the buck's unrest.

Ungalla, in full view, was as still as a blackened tree trunk. A birdcall shrilled through the cool morning air, and the buck took one undecided leap before continuing to feed.

Ungalla was so close that even his breathing was guarded. He drifted closer yet and paused within sure range—a scant twenty yards. The muscles in his spear-arm tensed, his wrist began to quiver silently, vibrating the shaft so that its flight would be true. He swayed back as though overbalancing. His left hand dropped, right arm and woomera flashed into one straight line, and the spear darted.

There came the muffled impact of the blow, and, struck squarely

through the lower part of the back, the victim dropped in the grass. The survivors, panic-stricken, crashed blindly through the bush and thudded into the distance.

The youthful Milunga must have disobeyed his senior's instruction to remain far in the rear as it was he who reached the dying animal first, dashing past Ungalla as that startled warrior whirled round, immediately on the defensive, at the sound of running feet. Nor was there any need for haste in this instance, for the quarry had been fatally wounded by Ungalla's broad-bladed killing spear. However, the kangaroo's dying struggles were now quickly ended by a blow from the young warrior's club; and Milunga then hastened to recover Ungalla's woomera, which had fallen to the ground. He scrutinised it carefully, licked the ball of his hand, polished a scratch—probably visible only to his eyes—and courteously returned it to Ungalla.

Ungalla said nothing. He had been about to reprimand Milunga but, after all, it was apparent that the young man had been skilful enough to follow without scaring the game and, as Ungalla was wryly aware, without the knowledge of the more experienced hunter. So the older man, absolutely expressionless, merely squatted down beside the dead kangaroo and using the quartzite chip set in the handle of the woomera as a knife, made a small incision in the centre of its chest. Thrusting in his hand, he drew out those edible morsels that belong by right to the hunter—the heart, the liver and the lungs. These he threw to one side and, again plunging in his scarlet arm as far as the elbow, dragged out the intestines.

Milunga handed him a short, hard twig. Ungalla shredded one end to a point with his teeth and skewered the cut so that the blood would be enclosed.

Milunga was rubbing the intestines in the dust. When Ungalla had straightened one of the kangaroo's long hind legs so that it lay parallel to the tail, his companion passed the intestines to him. With three or four turns of that gruesome rope Ungalla bound them in position. Then he pulled a forepaw down to the same point and encircled it. The other forepaw was then similarly tied, the kangaroo being forced into a more or less compact circle.

The hunter passed the end of the fastening under and jerked it tight. Then, with Milunga's aid he hoisted up the prize, set it upon his head, and made back to camp at a steady jog-trot; his companion brought up the rear with the spears and woomeras.

Tirelessly the leader swung through the sunlit bush. The air was still cool, but sweat gleamed on his black hide and ran in little rivulets down his spine. He strained up the ridges and jounced rapidly down the slopes, the thick pads of calloused skin on his feet protecting him from the thorns and jagged flints in his path.

Suddenly a wild yell rang out, then the rattle of lubras' voices and the shrill cries of children. Half a dozen dingoes, each leaving a wake of golden dust, raced hopefully out to meet the hunters, and naked black bodies showed through breaks in the bush ahead.

Ungalla, basking in adulation, stalked into the small clearing he had left a few hours earlier, and flung down his kill.

Fortunately for their appetites the tribe was noisier than it was numerous. Yet there were enough to make the preparation of the food easy.

Milunga and two of the old men dug a shallow pit, flung in dead branches and lit them. Ungalla unbound the kangaroo, sliced the skin of the lower leg with his woomera, and exposed the great sinews. He pushed a strong stick beneath one of them, steadied the animal's body with his foot, and twisted the stick powerfully. The sinew tore from its anchorage near the top of the limb, and, severing the lower end, he looped it carefully and tucked the precious silvery riband into his dense mop of hair. He drew two sinews from each leg, then nicked the middle joints and broke them. Milunga hacked off the tail and with Ungalla's help held the kangaroo over the flames to singe.

The now high-riding sun lit the scene with a fierce brilliance. Heat waves danced above the surrounding bush. Runnels of sweat cut little paths through the dust sticking to the workers' skins. As they scraped the charred fur from the kangaroo they shielded their faces from the shallow, fiery pit.

Ungalla raked a hole in the coals, and the carcase, now weirdly misshapen, was lowered in. They threw the tail alongside and heaped the embers over all.

Ungalla wiped the perspiration from his forehead; then he and Milunga squatted close to the fire and rolled the insides of the kangaroo in the coals. Almost before they were properly heated, they scraped them out with sticks and began to eat with slow enjoyment.

A young child, scarcely walking, seized a fragment and bore it off in triumph, but otherwise these prized morsels were eaten solely by the hunters.

While the odour of burning flesh became stronger, Ungalla and Milunga, wiping the blood from their mouths, acted the stalking and killing of the game.

11

Milunga mimicked the kangaroo, alternately feeding and glancing up suspiciously while Ungalla stole towards him. A shriek of appreciation greeted Ungalla's deadly throw, and the applause grew louder as Milunga leaped convulsively and keeled over, pawing feebly at the dust.

But soon the smouldering pit was again the target of every eye.

Ungalla dragged out the partly-cooked kangaroo and dumped it on a bed of green boughs. Deftly he sliced the stringy meat, dividing it amongst the men.

Milunga got part of one hindquarter, and the medicine man got the butt end of the tail. Ungalla would often pause to sever a fancied piece for himself. He passed a foreleg to each of two ancients and gave the rest of the tail to a younger warrior.

The lubras and the children eyed the succulent chunks hungrily, but their turn was not yet to come.

With only the headless, limbless trunk left, Ungalla motioned to his companion of the chase. They tipped the body so that the entrapped blood ran to one side of the body cavity; then, tilting it still further, each drank in turn.

That was an important part of the feast, for the blood had great virtue and bringers of red meat must be kept strong, cunning and brave.

Ungalla cut the body into five parts, which he handed to the younger men. When each man had cut or chewed or ripped whatever he wanted from his share, he handed the remainder to a lubra. The lubra handed what was left to the children. The babies got the bones or, if they were specially indulged, a chunk of gristle to chew.

Yet, one thing was certain. When all the niceties of custom had been observed there was not an unfilled mouth in the company.

As the sun grew hotter the camp drowsed contentedly. Only the ever-hungry dingoes prowled through the blazing heat of noon. If they had not eaten well, at least they had eaten as well as they could.

And that, too, was as it should be, for every Aborigine knows that a fat dog is a lazy dog, and a lazy dog is useless.

# The Ocean Wanderers

IT was as though there were two oceans—an ocean of water and an ocean of cloud. Nor was there any living creature to be seen, either in the air or on the sea; nor any movement other than the long steady march of billows, smooth but unquiet and so unvarying that the desolation of that primeval scene was intensified rather than relieved.

Then a storm, driven by a strong wind, trailing dark curtains of rain and with lightning glimmering amongst its sombre steeps and chasms, swept out of the east, and circling before it were the birds— black, long-winged petrels that wheeled and swung and soared through the dim afternoon. Sometimes, so closely would many of the flyers skim along the troughs of the waves, that the company would seem suddenly to decrease in numbers or, indeed, to vanish entirely; and then again the sky would be aflicker with speeding shapes rising into the wind to turn and swoop again down the long, moving valleys of the sea. They seemed to be flying for the love of flying. As they swung back and forth there was seldom any movement of their wings though their progress was as swift as it was effortless.

It was late when the petrels began to settle. Low in the west the sky had cleared and the setting sun had turned the ocean from grey to blue. Also, the colour of the petrels was now a deep reddish-brown, and not black as in the gloom of the rainstorm. They were rather larger than pigeons, with webbed feet and black bills, strongly hooked. Their bills were their most distinctive feature, the nostrils being produced into tubes which lay along the upper mandible and gave it a thick appearance near its base.

The birds clustered together while the rim of the world blotted out the sun. They made no attempt to fly landwards, for as doves are at home on the rooftops and sparrows under the eaves, so were the pet-

13

rels content in their own unstable element. Stars shone from the black skies, the waves glinted mysteriously as they swung into the dark and, after the breeze had died, there was no sound but the occasional splash of a wave's crest and, at intervals, an outburst of shrill cries as the petrels fought over some scrap of food. And these squabbles were the more incongruous because of the serene night and the nets of silver the rising moon had cast over the waters.

By dawn most of the petrels were in the air again. They circled awhile over their resting companions, then sailed on in search of new feeding grounds. At times, they flew so low and so slowly, almost hovering, that they would suddenly run for a short distance along the surface of the waves with a loud, pattering sound, their wings waving gently, their weight mainly supported by their webbed feet.

Before the sun had cleared the horizon all the flock had settled on the sea again to feed on plankton—those innumerable transparent living things which are found at varying depths and in varying concentrations in every ocean of the world. Often a number of petrels would dive in pursuit of small fish preying on the plankton from below, the birds using wings and webs as they darted and turned and twisted through the clear water, to emerge at last, with a rush, back into the sunlight, riding high on the shimmering skin of the ocean, their buoyancy making the ease of their descents the more remarkable, for most adept divers float low in the water with only the head and neck visible.

Day after day, week after week, flying and resting, in sunshine and in rain, in calm or boisterous weather the birds lived contentedly in their world of wind and sea. However fiercely the gales blew, the long wings of the petrels rode out the gusts; and if a storm were moving across the ocean they would sometimes play about its edges, swirling through the margins of the downpour or wheeling into the mist of rain itself, as black as bats. In bad weather, even when sharply defined in a lightning flash, it was easy to fancy something of the supernatural about them. They were so swift and silent, so eerily active and unconcerned in the midst of so much noise and violence that it seemed they were not birds at all, but the ghosts of birds.

One such rainstorm which came in the early afternoon of a cloudy day was followed by another and then, suddenly, the ocean was savaged by a furious wind. In a moment the wide, cloud-darkened scene was brightened by distances of foam, and the noise of the waves as they broke and flew under the pressure of the gale was like the roaring

14

of surf against cliffs. Of necessity, every ocean bird was a-wing, for none could have floated in the white turmoil below.

The petrels sailed higher and more swiftly than usual. Where they had followed an elliptical path before in journeyings that were sedate, indeed, compared to their present flight-patterns, they now raced down the sky in wild, sweeping arcs that brought them close to the water at one moment and then sent them rocketing skyward at such a speed and to such a height that they often seemed to be moving through the fringes of the clouds.

For three days and three nights they were unable to alight and, when at last they did settle, there was no food, for the plankton had gone deeper down. So the weary flock took wing again and ranged over the face of the ocean as before, but still without success, though the waves were dying and the wind had ceased.

That afternoon it began to blow again. Deep green clouds welled out of the sea; lightning split the approaching storm; thunder drowned the crying of the birds and again the wind clawed the ocean into shreds. All through the clamorous night the petrels spun and circled in the gusts as though they were not subject to the ordinary fatigues of flesh and blood, and it was not until next day that the starvelings began to fall. Drowning was swift in that smother of white water. Dawn showed a great company still; but that evening and through the night the petrels were blown out of the sky in increasing numbers. Only the strongest survived and almost half the flock was destroyed before the second gale passed. It was a weary crowd of survivors which fluttered down at length to rest upon the water. They settled while the waves were still mountainous and, swinging dizzily in the huge seas, drowsed with drooping wings. More dead than alive, they began to feed next morning when the waves had diminished and the plankton had moved out of the depths—and the sun remembered to be warm again.

About a month later, the petrels came in sight of land. There appeared to be two islets, close together; but as the flock slowly approached these were revealed as twin peaks with a common base. Even in fine weather there were often pennants of cloud flying from these peaks, generally in a northerly direction. The main part of the island also trailed northward in the same direction as the cloud that so frequently shaded it. Low-lying for the most part, it was covered unevenly with forest.

Patches of kentia palms gleamed a shiny greyish-green here and

15

there, especially near the edge of the sea and there were clusters of pandanus, a few banyans and some open, fairly level stretches of grass.

As the nomadic colony—still great in numbers—approached this speck of earth in a wilderness of water so did the casual group life of the birds change and they divided into excited clusters that became denser and denser, crowding together and often settling on the water with loud, harsh cries and frantic splashings and flutterings. Each afternoon for three days the petrels carried out the same extraordinary splashing ritual and then a few individual birds detached themselves from the rest and flew purposefully landward—crossing the narrow strip of beach and circling over the grassy slopes and the palms and pandanus. They did not land, however, though their sweeps above the grass and the trees became shorter and shorter until many of them were merely fluttering. They would hover strenuously for a while, then swing away in sweeping curves that took them over the familiar sea again, only to turn into the light wind and hover as before above the place where they had paused a few moments earlier. Each time they returned they became bolder, until at last they were drifting back and forth just clear of the trees.

At dusk a few petrels started to land by a forest of palms and with flapping wings and awkward, hopping gait ran swiftly between the gloomy trunks, rustling over the brittle, dead leaves. More followed and soon the place was resounding with the cries and the blunderings of many birds searching for nesting sites.

From then on the voyagers arrived in increasing numbers, settling in the grass on the slopes above the beach and at the borders of the wooded places, each bird hurrying off in one direction or another upon alighting, so that the scene was one of breathless and confused haste.

By the time it was dark the sky was raining birds. All along the length of the eastern side they swept in from the sea in hundreds and, indeed, in thousands, hovering briefly and then descending to the earth they had not touched for so long. They landed above the middle beach and on the steeper slopes and forested ridges on each side; and if there were no suitable landing place, as on a small headland packed with palms, they would fold their wings and fall blindly through, to thump heavily to the ground which, fortunately for them, was some- what resilient because of the massed fibrous tree roots. Everywhere, all along the eastern shore the petrels were coming to earth.

That night, after months in mid-ocean, the tang of salt gave way to scents of moist sand and dead leaves mouldering in drifts between the exposed roots of the trees and in the entrances to the many nesting burrows. Instead of the splash of water there were the silken rustlings of palm leaves and the murmuring foliage of a great banyan. On clear nights there had never been anything to hide the stars, but now the swaying kentias and pandanus waved long black fingers overhead and instead of the heaving waters of the ocean the petrels had the solid earth to support them.

Later that night, when the grassy slopes stretching down to the beach were frosted with the white glare of the moon, the petrels began to repair the nesting sites of last season, shooting out the sand with powerful thrusts of their webbed feet as they re-tunnelled burrows that had caved in and cleared the entrances of leaves, twigs and other debris. Next morning there was no bird to be seen when dawn at last revealed the island and the wide, grey sea. While it had still been dark they had run silently from the trees, from the clear places on the fringe of the thickets, and from the grassy slopes to gain height. Crossing the pale line of surf, they were again skimming the ocean, back and forth, as they had done so often before.

The cleaning and repairing of old tunnels and the making of new ones went on for the next five or six weeks, then the large, white eggs were laid, one to each mated pair. Each day, in every kind of weather, half the flock would fly out to sea while the other half remained in the nests—to be relieved about dark by their partners. So at dawn and dusk every day there was now a double movement of the petrels, with hundreds coming in to the island and hundreds leaving for the sea.

Steadily, unhurriedly, neither early nor late, the cycle continued. Some of the eggs were taken by predators in the short intervals when they were unguarded but, by mid-summer, there was a fledgling in most of the burrows, with both parents now working strenuously to keep it supplied with food. Each evening the warm dark would be filled with the cries of the adult petrels and the shrill, subterranean clamour of young. Issuing strangely through the sandy loam these cheepings and chatterings would continue until the incoming fishers had scurried into the pitch darkness of their nesting tunnels and yielded up the larger part of the day's catch to the insistent and voracious fledglings.

Nothing, it seemed, could upset the breeding and rearing of petrels. Even the tropical storms that swept the island failed to disturb them for the heaviest rain simply soaked into the sand and away so quickly that the occupants of the burrows seldom suffered more than a wetting.

As the weeks went on the single chicks squatting in the tunnels became larger and heavier, and, as they grew older, they ate more. Their parents fed them tirelessly, until they themselves were lean from constant food-gathering and the chicks were paunchy, overfat, downy caricatures of nestlings. Insatiable though they were they never moved from the place where they had come into the world. They sat stolidly all day long, staring into the dark or sleeping for hours on end. The only animation they ever showed was in demonstrations of eagerness at the prospect of more food.

But the day came when some of the adult birds did not return; on the following night, a yet larger number remained on the ocean. Soon there were no birds flying anywhere in the vicinity; the burrows were silent, the skies empty. The nomads had passed on.

The deserted nestlings remained in their tunnels. Gradually, the down that had covered them fell away and feathers began to grow and harden. Occasionally, a young bird would emerge and sit awhile on the cushion of sand which had been scraped from the tunnel at the start of the breeding season and, as the days turned into weeks, so did the bloated shapelessness of the new generation shrink to the slim lines of flyers that could range far and tirelessly.

One calm morning in autumn—unexpectedly and apparently without any feeling of apprehension — some of the young petrels ran unsteadily down the island's eastern slopes and, flapping strenuously, cleared the beach and made out to sea. More of them left the islet in the next few days, until every one was a-wing, and it was not long before they, too, had learned to sail on motionless pinions so that there was soon a similarity between the flight patterns of the younger birds and the surer, wider curves of the parent flock now far away from the rookery which had been the starting point of all their lives.

# The Assassins

AS the dingoes crossed the exposed crest of a ridge the wind raised small ruffs in the longer hair of neck and shoulder and whipped the grey dust from beneath their paws. Then they descended into the hot stillness of a gully, silently, keeping the same pace, weaving between the brown grass tussocks on the slope and skirting the embedded boulders where the heat waves danced.

They were a mated pair, a red dog and a tawny-coloured bitch, and they had been following the cattle pad since dawn. It was now midday. They ran without haste. The dog usually led, for the way was narrow, but occasionally, where the going was easier, the two would travel abreast. They were beautiful with their sleek coats and bushy tails, their broad, intelligent heads and compact, powerful frames.

There was a level, grassy patch at the bottom of the descent, a sheltered, park-like place which held a tinge of green and where, to heighten further an illusion of peace, a black bull drowsed in the shade. Grumbling deep in his throat the huge creature pivoted to face the newcomers. The dingoes veered away and circled, making an almost inaudible rustling in the grass. They kept their distance for a while before the dog darted in. But the bull seemed reluctant to become involved and made no forward movement at all, not even when his horns lunged up in that terrible, lifting arc which can disembowel an enemy at a stroke.

Persistently, cleverly, the dingoes harried him until at last a short but purposeful charge brought him into the sun. And he was old in an instant, his gaunt frame notched with hollows, his black hide scribbled over with the record of his battles. Before, he had loomed larger than life; now his sinewy, scraggy neck mocked the muscled column of earlier years, and the network of veins which snaked along his flanks was suddenly conspicuous.

Of former greatness only scars remained. For weeks after his last and worst defeat he had hung about the fringes of the mob, bellowing a challenge he could no longer make good; then, gradually, he had lagged in the strenuous journeyings over that wild terrain, finally to lose touch altogether with his kind. The days of his strength had gone.

But age had not dulled his understanding, and he would not be goaded into another charge. He would expend no more of his strength.

Suddenly turning away, the red dog trotted to a nearby patch of shade and flung himself down, panting. The tawny dingo also ceased her running backwards and forwards and came to a halt a short distance in front of her huge opponent.

The black bull lowered his head and the deep-toned vibrations of his displeasure became stronger. His hoof ripped a dusty furrow in the grass. Drooping abjectly before him the dingo bitch sidled nearer, then sat back on her haunches. Her white fangs glistened in the sunlight and her tongue was as red as a canna flower. Hollowing her back she slid out her forepaws until they brought her belly-flat to the ground where she lay with her outstretched neck resting comfortably across her front legs, her narrow eyes fixed unwinkingly upon the black bull. Her yellowish form had flattened itself, and was utterly still except for a twitching of the bushy tail.

One hesitant step at a time the bull rocked closer, his unease momentarily swallowed up in curiosity. Not until his square black muzzle was almost directly above the broad, sleek head did he come to a stop, tense and puzzled. The gently waving brush floated slowly down into the grass and there was no suggestion of life in the dingo—apart from the brilliance of her watchful, crafty eyes.

For perhaps five long seconds the tableau remained undisturbed. It was broken then by such an explosive leap and such a vicious snap of heavy jaws, that the bull was startled into an absurd backward jump of alarm. Next instant, he was thundering in pursuit of a bushy tail dancing golden in the sun and always out of reach. Repeatedly, it whisked just under his nose, drifted sideways, or sped away in front of him while the grass clods flew and the powdery earth spun into dust under the violence of the black bull's turning and twisting.

Always the yellow dingo kept one small jump ahead. Eagerly, her red-brown mate joined in, each covering the other, working perfectly together as they crossed each other's tracks, or dashed in, or circled, or simply melted from the path of disaster. It was as though each massive lunge of the bull's horns puffed aside the target as the stroke of a bludgeon wafts aside the thistledown, and soon the movements of the great animal became more uncertain and clumsy. They baited him until the foam dripped from his muzzle and, at last, with pumping flanks, he slid to a halt.

The pounding of hooves and the dart of lithe shapes through the grass stopped. The dust drifted away and in that hard, brilliant light the three animals were so emphatically limned that they might have been cut from coloured cardboard.

The yellow dingo was first to move. She began to swing restlessly back and forth, back and forth, raising a tenuous thread of dust each time she crossed the narrow cattlepad. For a while the bull stood his ground, sometimes giving a defiant shake of his horns but generally showing signs of a mounting disquiet. Then, pivoting heavily, he began an unwilling retreat, his hooves clicking on the hard-baked earth. As he left the shelter of the gully dust fluttered from under his hooves like flocks of grey birds and the grass tussocks bent in silvery waves into the distance.

Mostly, the wild dogs kept well to the rear. Though they had been on the move since dawn there was no change in their gait even on those occasions when the strongest gusts caused the bull to veer momentarily off-course. Their patience was limitless. They seemed content

21

to follow for ever in their smooth and leisurely manner. But as soon as their quarry showed any inclination to pause they would close in and begin to harass him until, for the sake of peace, he went on. With hunting craft as old as the eroded boulders around them they drove him through the mid-summer afternoon.

Slowly, the shadows of the three became longer; slowly, the eastern sides of the ridges darkened and assumed a faint bluish tinge; the wind ceased to savage them as fiercely as before and there were times when a strange, unreal quiet overspread the scene.

Shortly before sundown the cattlepad dived into a tangle of scrub and fallen timber where it unravelled into many paths which led by different ways to the foot of a red earth cliff. The place was full of shadows. Raising his head the fugitive scented the quiet air, then broke into a canter, sweeping aside the bushes with his heavy shoulders. The dingoes followed, silent and almost invisible. Bursting suddenly into the open the black bull stopped. The waterhole where he had camped for many a night in the past had almost disappeared and the scent of wild cattle was stale; pitted hoofmarks in the mud were dry and hard and nothing moved in that desolate place except a few parakeets, fluttering and chattering on a dead branch at the edge of a pool of green and stagnant water.

The black bull stood awhile. He dropped his muzzle to sniff at the earth, then went forward again towards the water. Though neither fresh nor sweet the smell of it was cool in his nostrils and he had travelled far without drinking. One of his forehooves punched a hole in the dried surface crust, but he went on, then broke through again, frightening the parakeets by his plunging efforts to free himself. He turned back only when he saw the dingoes racing lightly towards him. But he managed to regain firm ground unscathed, and the engagement was still no more than feint and flourish, threat and counter-threat, with no bruise sustained and no blood spilt. For all that, there was death in the air as the black bull plodded through the scrub towards the red west. They drove him mercilessly on. Sometimes, he turned to fight but they would not close.

Another two hours had passed and the high hills were as pale as ice in the moonlight when the bull reached the brink of a short but steep descent. Again he turned to face his enemies, but they slunk back into the confusing misty brightness. He nosed at the slope. Loose stones rolled from under his hooves, and as he picked his uncertain way down the dingoes reappeared. They glided round him like leopards, seeming to acquire strength as their quarry weakened. Just as each mock attack throughout the day had been effective in its own way, so was the death-stroke adroitly carried out.

Darting in from one side the dog slashed at the bull's shoulder. Simultaneously, and from the other side, his partner flung herself at the quarry's undefended flank. Driven by all the weight of that compact, muscled body her jaws struck low down and in front of the hind leg where the hide was thinnest. There they sank so deep that the terrible, shearing bite literally tore out a gobbet of flesh and the force of the impact, together with the shock of the wound itself, brought her victim to his knees. Nor could he rise at once and when at last he did struggle to his feet his flank was wet and glittering. Blood had gushed over the ground where he had fallen and as he careered wildly down the slope it splashed the dry grass along the way. The big vein which coursed along the underside of his belly had been severed and there was little likelihood of the flow ceasing as the doomed creature lumbered strenuously on. His breathing became louder, his stumblings more frequent. Behind him, smoothly, ran the killers. In the pallid glare of the moon the yellow dingo, drifting on as silently as a ghost, looked larger than her darker-coloured mate. He was a shadow but she seemed to glow in the moonlight. They made no further attack—as though they knew they had only to wait, for it is seldom that even a creature in its full strength is able to recover from the bite of the dingo.

The bull was easy to drive now. Every time he glimpsed one of the smoothly-gliding shapes he would make a strong effort to escape, forcing himself along with what strength was left to him.

He weakened suddenly. Making a run at a slight rise he staggered on reaching the top, then fell to his knees. The dingoes, too, came to a halt. After some moments of stillness, the huge animal rose and went on. But the end was very near, the progress of the three diminishing, then ceasing altogether. Propped on wide-spread legs the black bull nosed blindly at the earth as though trying to scent out the path he would take. He began to tremble violently and dropped to his knees. Once down, however, he seemed to lose his fear, probably because his senses—even the sense of smell—were now dulled by approaching death. For the first time since the start of his flight he was oblivious of his enemies and lay as quietly as one of his domesticated kinsmen resting in a moonlit paddock close by a homestead. Slowly, the sombre head dropped to meet its shadow on the ground and when at last the black bull had ceased to move the dingoes trotted in to claim their kill.

# *The Silver Forest*

STANDING as it did upon high ground, the tree was quite the most conspicuous object thereabouts, its loftiest limb, dead and bleached a silvery white, drawing the eye by reason of its upward and outward sweep. And about half-way along that soaring curve of wood, in the full glare of the moon, was the dark, distinct shape of a possum glider—a creature about the size of a cat.

Few native animals would have been so indiscreet as to remain for so long in so conspicuous a place, but this individualist seemed serenely at ease as he twisted into one complicated position after another in order to scratch each part of his body.

There was no wind that night, and even the whistling of teal passing overhead was clearly audible as were many other cries and call-notes. There was also a more unusual sound, not loud but often repeated, now close, now at a distance—a sound like the single clap of cupped hands.

None of these evidences of life, however, appeared to interest the solitary watcher balanced at such a precarious height above the gully; but he paused briefly in the combing of his fur when there emerged from a hollow at the base of the dead limb a counterpart of himself in all respects except colour, which in the newcomer was almost white.

The darker of the pair now ran to the end of the branch to peer down into the restless shadow nets entangled in the trees below.

He crouched and, with a startling cry, leapt outward and fell. Down, headlong, then smoothly flying in a magnificent, curving swoop with legs and gliding membranes at full stretch and long tail a-stream. He skimmed the trees in the gully and gradually slowed, floating through an opening in the branches. Faster now, swerving round a great, pale column, down through the patchwork twilight almost to the ground, he rose finally in a gentle arc to a smooth-barked trunk already scarred by the marks of many landings. And the end of that planing flight was marked by an impact as quiet and as undisturbing as the muffled clap of hands.

From behind and above came another screaming call, louder and steadily louder, then changing to a low chuckling as the white glider slid expertly through the branches overhead and alighted with the same distinctive slap close to her companion.

Clawing his way higher in a succession of convulsive, scrambling jumps the leader clutched at a ribbon of bark trailing beside the trunk and swarmed up nimbly. Having the same flair for rope-climbing, his mate followed at the same speed, in no way inconvenienced by the giddy swaying of the bark. Once close to the tree's crest they launched again into the swimming deeps of moonlight, travelling for the time being in a series of short volplanes until, from the top of the next ridge, they again swept away in a dive that took them down to the forest floor.

Here the gliders parted company, the female continuing on her way while the darker of the pair climbed to the top of a manna gum, and having reached its topmost plume, heavy with blossom, clung to the yielding, swaying branchlets, his long tail waving from the mass of foliage and flowers he had drawn together for support. Broken blooms rained down, twisting and turning past the stronger limbs underneath, and all the time was the subdued rustling of his movements with an occasional heavier swish of leaves as he changed his position. Once he descended in a slow, floating leap to a lower cluster which gave so flexibly to his weight that he was left swinging upside down like some big bird, but never ceasing to draw the blossoms towards his mouth. He fed unhurriedly then dropped to a firmer foothold.

As he searched out his path to the next landing a small shape came drifting through the tree-top and a miniature of himself alighted soundlessly on the same branch and darted in to nose at him in elfin curiosity. The next moment the tiny sugar glider had gone, sinking so quietly into the whispering ocean of leaves that it seemed questionable whether he had been warm flesh and blood or merely another trick of the moonlight.

The dark glider continued his journey, often uttering his shrill cry as he leapt, but making each climb to a high point in silence. The track he followed was well used, the marks of many claws criss-crossing the landing places, and where there was no bark-trailer to make climbing easy the trunk of that tree would be scored from bole to lofty branches. Even on the longest leap of all, where he fell far short of a grey gum, there was a boulder rubbed shiny on top and a narrow but unmistakable trail through the scanty grass and mould. Down on the alien ground the glider moved clumsily as though literally falling over himself in his anxiety to leave behind the spiky grass tussocks that might conceal an enemy.

Yet there were perils in the air as well and it was when he was sailing into the revealing brightness of an open space that a loud, harsh cry caused the small aeronaut to waver agitatedly in mid-flight. The trees on the far side were out of reach and he was able to plane only as far as a dead stump about three-quarters of the way across the clearing. It was a well-used landing place and from there a quick rush across the remaining stretch of open ground had always brought him to safety. But with the hunting cry of the most ruthless destroyer of his kind echoing in his ears, he stopped where he had landed.

Above the fringing trees where the moonlight sparkled a great bird beat slowly into view and floated directly towards the glider's inadequate refuge. The hunter—a brush owl of the coastal rainforests and scrublands—was a giant, even of its kind. Almost as powerful as the wedge-tailed eagle of the southern Alps, the owl, for all its size, might have been a wisp of fog, so gently and silently did it waft down. Nor was there yet the faintest whisper of a sound, and the illusion of intangibility was lost only when the owl's talons rasped on the weathered stump. Almost at once the great bird again rose into the air and, with a speed that was magical in so heavy a creature, swept into the darkness of the forest.

Still, the glider waited, not until he judged the marauder to be at a safe distance, but until he had forgotten the reason for pausing. Slithering awkwardly to the ground he then set off in a succession of noisy, scrambling leaps and so regained his familiar territory of trees.

By this road he reached the summit of the next ridge and planed to a manna gum. Immediately two sugar gliders landed in quick succession, darting up to the stranger to touch him with their inquisitive noses before whisking about and skimming away into the labyrinth of light and shade.

The dark glider began to investigate each of the innumerable tiny pits in the smooth bark, for the trunk was scarred from top to bottom and on every side. Clinging closely with his gliding membranes spread and his long black tail waving, he licked greedily at the sweet gum oozing from the wounds in the bark. Often his sharp lower teeth cut deeper still in quest of sap and whether it was done intentionally or not, those fresh incisions would bring a new flow for the next night's visit.

His white companion had reached the feast earlier and the pair systematically worked over the trunk and along the larger branches, giving all their attention to the pleasurable business of finding the palatable nodules and chewing further into the sapwood.

Often they would hang head down, and indeed they assumed every imaginable position, feeding in silence except for the brisk scratching of claws as one or other of them passed on to the next spring seeping from that inexhaustible, vertical stream.

It was almost dawn when they began the return journey. The moon had set but darkness was no hindrance and their progress was steady. Two more gliders now attached themselves to the first pair, the four travelling in procession. One turned aside, having reached his home tree, while the others continued to follow the leader.

With the clearing uneventfully behind them and the tree that was their goal already visible they swung in file along the last slope and were all in mid-flight when, for the second time that night, the cry of an owl sounded close by. Instantly, the dark glider fell to the ground. The other two planed on towards the tree that was their immediate destination.

It was the white glider's extreme good fortune that there was now another glider behind her, for the owl had caught sight of the voyagers, and darting through the branches with the speed and deftness of a falcon swept up the hindmost so silently that the death of the small victim was almost secondary to the eeriness of the incident.

There was a definite lightening in the east before the surviving pair ventured back to the higher branches to launch themselves on the final leap for the base of the dead tree on the ridge.

Two small, quick shapes came flying through the first gleam of day, there were the almost inaudible impacts of two landings, a rattle of claws on the tree's hard shell, and then only the smooth branch shining in the early sun.

29

# The Marauders

HIGH above the headland's crest where the grey-green scrub whipped in the nor'easter a kestrel poised, keeping her position to the width of a flight feather, balancing in the tumult of the air. She waited, then sped downwind in a curve that took her low over the trembling bushes. Losing speed as she rose again, she wheeled lazily into the wind. She dropped and checked, wings beating so slightly that their action was hardly more than a controlled vibration of the tips. Directly below was a fairly wide patch of dry grass and it was upon this that the kestrel, her head thrust alertly down, was now intent.

The capture was perfectly timed. Plunging into the calm of the clearing she intercepted a big green and yellow grasshopper — a scanty reward for so much effort. She caught several of the slow-flying insects as they whirred like small, mechanical toys above the tussocks and then, tiring of such unsatisfactory fare, soared back to the top of the headland and began to work down the windward side, where the sound of the sea and the wind was as loud and as continuous as the roar of a waterfall.

Here and there a rocky outcrop gave some shelter from the blast and it was above these and above a few clear places in the scrub that the kestrel often hovered. She descended to investigate a flicker of movement at the bottom of a shallow gully but found only a swarm of tiny blue butterflies whirling over a mud patch. It was a place sheltered from the wind and she alighted on a dead twig, ruffled her feathers, and settled down to rest for a while in the warmth.

All the brightness in the world seemed to be pouring into that quiet hollow. Sunlight glittered blindingly on a water film seeping over a rock and heated the mud patch underneath until it steamed.

While the kestrel drowsed the butterflies whirled and spun bewilderingly. Except for them there was no movement, for the restless bushes at the top of the slope were not of the gully but of the headland, where the wind swept in unhindered from the sea.

Every now and then one of the butterflies drifted softly down to float just above the mud-patch and then to settle on its glistening surface. Blue wings would pulsate slowly once, twice, then fold together to show only their silvery undersides; and the numbers of those resting on the mud steadily increased until they looked like a scattering of greyish leaves.

If there was no other movement in the hollow, neither was there any sound but the sound of the wind, and that was so constant that it seemed scarcely to disturb the hush. So the sharp rustling which preceded the arrival of a newcomer was extraordinarily distinct. Yet it was not a loud noise, and the kestrel gave no sign of hearing it.

Emerging from the brittle dead stuff the lizard darted silently across the rock and stopped—no other creature in the world can stop as a lizard stops, or flash so instantly into action. When there is urgent business afoot he has no medium pace but moves in sections of speed divided by halts of such suddenness that the gaze skids ahead and must retrace its path to recover the subject.

From the tip of his triangular, reptilian head to the end of his tapered tail he was not much longer than a man's fore-finger but there was a deft and savage grace about him. He was so slim and sleek; he had appeared so dramatically, and stopped at gaze so surely and confidently that there remained with him even in repose an air of menace which was nonetheless real because of his small size. Basking there on the heated sandstone he shone in the sun with the brilliance of a metal ornament, each close-lapped scale so finely chased and polished that he glittered in iridescent gold and green and purple, yet with a magnificence so Lilliputian that from a little distance he was invisible against the dark rock.

He remained motionless until a twig broken off by the wind came spinning down to frighten him with its shadow. Shadows to him meant wings and he flashed over the edge of the rock, then quickly reappeared, moving easily down the overhanging sandstone face and rustling again through the fallen leaves on the ground.

It was from there that he caught sight of the company of blue butterflies and, craning up his snake-like neck and head, stared fixedly at the swirl of wings. Tense, alert, he stole forward, each movement

32

now a marvel of care. There was a smoothness about the slow reaching forward of each leg in turn which suggested that he was being driven by a mechanism of synchronised and immaculate precision.

Pausing, he seemed momentarily at a loss. Then one of the flyers came gyrating down to settle; the blue wings closed and opened once, twice, and then folded. No sooner had the lizard marked his prey, than his hesitancy vanished and he began to move forward, step by cautious step until the distance separating him from his quarry was little more than the length of his own body. There he paused to gather his legs under him; but the final rush—it was almost a spring—was too swift for the eye to follow in detail. First a shimmering streak of brilliance, check, and the small marauder with the blue wings of his victim quivering between his jaws, was whisking about and darting back to cover.

In a surprisingly short time he was back again and kill followed kill until his voracity appeared more remarkable than his prowess. Not every sally was successful, however, though it was rarely that he blundered, and his few failures were nearly always due to the same kind of mischance. Sometimes as he was stalking the butterfly he had marked spiralling down to its landing place, he would disturb another and, distracted by the shimmer of wings rising from almost under his

jaws, would lose sight of his original quarry. In such cases he generally moved a little further on, but in an aimless and puzzled manner, before returning again to the shelter of the dense bush serving him as a base. But he was never in hiding for long, and as soon as the next flyer alighted he would re-appear, as eager and as hungry as ever.

The kestrel had taken wing and, having risen high above the close-packed bushes, was just on the point of wheeling downwind when she glimpsed a flicker of movement amongst the leaves on the ground. She dived and checked, watching.

Then there occurred a coincidence that was eventually to extend the span of the lizard's life, for it chanced that a black falcon was then crossing the headland; it chanced also that the kestrel's sudden descent did not escape the sharp eyes of the larger bird and with a magnificent surge of power he hurtled towards the scene.

As soon as the lizard moved again the kestrel swooped, the falcon following, ready to filch the prize, whatever it might be. But when the biggest, strongest and swiftest of the three marauders did catch sight of the quarry he found himself balked by the kestrel. There was a glimmer through the dead leaves as the lizard took his chance of escape, and instantly the falcon turned furiously upon the smaller hawk. Deadliest killer of all in proportion to his size, he would undoubtedly have struck her down had not the kestrel plunged headlong into one of the dense bushes that grew profusely thereabouts.

It was by no means a dignified withdrawal from the field, but it was complete and thorough and at least the terrified fugitive, crouching dazed but unhurt in the leafy depths, was safely beyond reach of the falcon's ferocious temper.

Encounter, pursuit, and escape had happened within the time it would take for a single wave to break on the rocks below, but many waves were to spend themselves before the kestrel stirred.

When at last she did emerge the skies were again clear, and she was soon winging unconcernedly about her familiar hunting grounds.

White caps flared and faded against the deep blue distances and the wind boomed in steadily from the ocean, shaking the matted scrub and blustering across the wide, sheltered hollow where the blue butterflies danced.

Later, when it was almost dusk the little kestrel swept round to the southern slopes of the headland and hovered there as before—above the plumed grass tussocks where the grasshoppers whirred up from time to time on their green, unskilful wings.

# Wings in the Dark

CLEAR-CUT, the distant hills stood black against the sky, and already the faint blue haze of evening was flooding the gullies. The river, where it broadened, turned from silver to a sheet of beaten brass, and along its banks the smooth trunks of the red gums flamed.

A pair of honeyeaters flashed through the sunset brilliance and, zigzagging low across the water, vanished into the trees. Making his last journey before nightfall, a kingfisher skimmed upstream, and, for that charmed interval between daylight and dark the bush was vibrant with bird calls.

Then the last glint of sunshine faded from the tree-tops, and the impish, dancing bats were there pirouetting in the dusk. The reflections in the river dimmed, purple stained the shadows near the bank, and, from a branch arching above the water, a bird suddenly opened his wings and drifted buoyantly to the ground. The descent was soundless. There was a brief rustling as the frogmouth picked an insect from the dead leaves at the base of a grass-tree; then he rose as silently as he had fallen, and regaining his former place, clapped his beak loudly, as though relishing the quality and flavour of his first catch for the night.

But he was not given to unnecessary noise or movement, and a moment later he had frozen into immobility. As he crouched along the branch—not across it as most birds do—he might have been a piece of bark or the stub of a broken-offshoot; the miracle of his concealment lay not in keeping under cover but in the imitation of a natural object, of posing in full view of passing enemies and yet remaining invisible.

The dusk deepened. The stillness was invaded by innumerable tiny sounds—the mechanical skirring of insects, the croaking of frogs in the muddy river margins, the murmurous lapping of ripples through

35

the reeds. As the greenish light above the western hills vanished the frogmouth's slit-eyes expanded into great sombre orbs.

His head turned abruptly. About midway between him and the point where the branch joined the trunk a ragged tangle of sticks was wedged into a horizontal fork and could have been mistaken for debris left by a flood. Leaving this untidy nest a second frogmouth fluttered along the branch towards her mate.

They were a grotesque pair and almost identical in appearance. Each was rather larger than a dove, with a heavy head from which the feathers swept back in the same manner as those of a kookaburra. But the beak was their strangest feature. It was absurdly wide, ridiculously stubby. Such a beak as a means of catching insects was ideal, and the margin of error it allowed its owner would counteract many an ill-judged snap. But it was hardly an ornament. It was, in brief, reminiscent of a frog's mouth.

Yet the misty softness of their feathers transformed the birds from the ugly to the quaint. The plumage was of so fine a texture that it seemed to have been dusted with grey powder, and its pattern was the pattern of crinkly bark mottled with lichen. There was no sheen on the feathers. By day, when the birds rested, they absorbed the light instead of reflecting it.

From the other side of the river came the cry of an owl, and one of the frogmouths flew to the nest. Instantly there was a vociferous chatter of eagerness. A single ray of moonlight, striking through the foliage, illumined the downy bodies and the gaping, impossible mouths of two nestlings.

The other adult bird winged away like a huge black moth. Passing silently between the trees he began the ever urgent search for food, sometimes swooping as sharply as a hawk, sometimes floating leisurely down to claim a slow-moving ground insect. When he paused to watch from a vantage point the habit of stillness was marked and he melted into invisibility. He never moved without good reason.

His mate joined him, and the two wafted restlessly from the tree-tops to earth and back again. Pale light and black shade flickered incessantly on their wings. They hunted noiselessly. There was a ghostly quality about the chequered moonlight, the whisper of leaves high above the pale-limbed gums, and the two birds fluttering through the dimness.

They made innumerable journeys to the nest. Each visit was marked by the same ravenous clatter from the nestlings, but generally the adult birds were mute. As might have been expected, when they did call the sound was unusual—a succession of monotonous but subdued booming notes.

As one of the full-grown birds was about to leave the nest on yet another excursion a piece of bark cracked sharply and fell. The frog-mouth whirled and crouched—a fearsome, nightmarish monster. She seemed to double in size, even the feathers on her head rising. She quivered, vibrating with rage, and, opening her fantastic beak to its fullest extent, uttered a stream of harsh croaking noises.

The object of this display, until now lying quietly against the branch, changed his appearance almost as completely. Except for the accident of the snapping sliver of bark he would have lain undetected until his opportunity came, for he was as much a master of the art of camouflage as the frogmouth. Indeed, allowing for the coldness of his reptilian nature, he had much the same character. Matching the threats of the bird with a sound like the hissing of a steam jet, the jew lizard dilated the pouch under his jaws to its fullest extent, opened his mouth, and advanced a step in a valiant attempt to terrify the bird into retreat.

But the other did not move. She covered the top of her nest as well as the nestlings, and the duel degenerated into a tableau. The lizard had no stomach for a frontal attack. He was less daring than his swash-buckling kinsman, the goanna, and he preferred robbery by stealth to robbery under arms. Even so, he was loath to abandon all idea of tak-ing the nestlings.

So he waited, unwilling to go on, unwilling to retreat. Moonlight rippled through the branches, and the delicate tracery of moving leaf-shadows blurred every outline into haziness.

Far away an owl was hooting, and the cry carried clearly through the bush. A fish jumped in the river below and multiple circles bal-looned towards the bank. But neither actor stirred; they stayed, facing each other, as opponents should.

A dark shape sank through the crest of the tree, almost reached the branch, and abruptly rose again. Glimpsing the belligerent attitude of his partner, the newcomer circled watchfully, peering down with wide, gloomy eyes. At last, he made out the unmoving form flattened along the branch.

The lizard flicked back like a clockwork toy as the frogmouth darted noiselessly past his snout. Looping, the bird dived again, and the metallic snap of his bill startled the reptile into another uncertain movement. His claws rasped on the wood and he fled, only to be attacked again as the original defender left the nest and joined in the onslaught. The loud clapping of their bills was incessant and the clamour became louder and apparently fiercer. The lizard retreated towards the main trunk, but the bright unfriendly moonlight, the height of the branch, and the speed of his enemies confused him. Thoroughly bewildered, he began to dodge, thrusting forward with jerking runs and undecided swings from side to side; yet the heaviest blow he had suffered was the whisk of a downy wing tip across his eyes.

The battle raged in make-believe until, exasperated to the point of rashness, the jew lizard reached up, snapping in blind fury, and his counter was more than successful. The movement was so swift that one of the frogmouths swooping to the attack failed to swerve in time.

There was a muffled impact, a croak of alarm, and two indistinct shapes spinning down towards the river.

After the first shock of surprise, the bird steadied, planed across the dangerous water and rose into the trees. The lizard, writhing like a snapped rope, disappeared in a burst of foam. But almost before the last drops had descended, an arrow-shaped ripple was pointing at the bank. There was a glimpse of a sinuous, gleaming body snaking out of the water and a crackling of leaves that became fainter as the river smoothed.

When the dawn had seeped through the crests of the highest trees and strengthened into the full light of day it showed only a pile of sticks on a great over-arching branch, and nearby, a small projection that could have been a piece of bark or the stub of a broken offshoot.

# *Paterfamilias*

A S the light strengthened, the splendid distances of the grasslands became more apparent, until at last the horizon was sharply evident, full circle. There was an impression of boundless space, just as there is in mid-ocean, and the impression was brought about in much the same way — by the vastness of the sky and, as a natural corollary, by the level nature of the plains and the absence of any focal point of interest. There was no outstanding feature in all the wide landscape unless the dark, meandering line that marked the course of the Murrumbidgee could be called outstanding.

Nor, so young was the day, was there yet any colour other than grey in its many tones and shadings, nor any breath of air to sway the grasses or to ruffle the reflections of the gums along the river bank. Stillness everywhere, or almost everywhere — for in one place where the river's breadth had been narrowed by a peninsula of flood debris, the current was fast enough to shiver the images of trunks and branches into sparks of black light and there a small, arrowshaped ripple was hurrying purposefully against the flow.

The sign was not easy to detect. It was made even more inconspicuous because it was moving into deeper shadow where the stream was strongest and the water most disturbed; and when at length the sleek, small form of a golden water-rat, shining wet, suddenly emerged there seemed to have been something mysterious, even magical, about his appearance. Picking his way carefully across the peninsula the solitary traveller, for there was no other living thing to be seen or heard, glided into a tangle of driftwood, came into view again at the top of the pile, and leapt from there to the trunk of a dead tree which had fallen, making a bridge from the low shore of the peninsula to the river's high northern bank. Completely disregarding the dangers of his exposed

position—a position most other native animals would certainly have avoided—the water rat at once set about the cleaning and combing of his fur, contorting his body in the process as though to display one by one and to their best advantage the many characteristics which distinguished him from his drab and infamous kinsmen of the cities.

For instance, he revealed that his hind paws were pale fawn, that his long tail was black with a white tip and that his belly fur was creamy yellow, shading to rich orange near his flanks. And, even if no living creature really merits the high-sounding title of 'golden', his darkish brown coat was certainly overlaid with a golden-tinted sheen that became more definite as the day brightened—altogether an individualist of brave and immaculate elegance!

Slowly, the first faint pink hues of the river red gums deepened. A pair of teal, flying low, rocketed downstream, skimming the silk-grey water. A magpie fluted from a branch high enough to catch the yellow sun and, higher still, small as an ant crawling over the translucent vault of the sky, a crow uttered its harsh, long-drawn call. A faint breeze rustled the leaves; colour blossomed; wild grasses quivered and nodded, and everywhere, on every side, day was bustlingly on time.

But the golden water-rat, though he had had no rest since sunset, showed no sign of quitting the stage. He was less noticeable than before, however, because he had ceased to move and seemed intent upon a particular spot above the low-lying peninsula. He continued to stare fixedly in the same direction until a red fox came padding through the bewildering shadow-patterns of the trees along the river bank.

Suddenly tense with excitement the watcher sat bolt upright, remained briefly at gaze, then, doubling around the curve of the dead tree, paused again with his fawn-coloured hind paws and the white tip of his tail in plain view. Eagerly—stumbling in his eagerness—the fox gave chase. As he sprang on to the dead tree he again glimpsed the fugitive, running up the sloping trunk ahead.

In spite of previous disappointments the fox was evidently young and hopeful enough to allow himself to be tricked into yet another enactment of this most unprofitable sequence—as though it had become the stylised conclusion to his every nocturnal excursion; and thus far, of course, each performance had ended in the same way. On this particular morning, too, the pursued gained the other side of the river the usual second or so in front of his pursuer, then swarmed down the vertical bank and vanished into a mass of roots which had been washed bare by the erosive action of the current.

Making no attempt to follow, the fox continued on his way without so much as a turn of his head, for there was at least one factor in this affair about which he had no illusions and that was the impregnability of the water rat's refuge. On the two occasions he had slithered down the bank to press home the attack his quarry had not bothered to withdraw very far into his gloomy labyrinth but had crouched unconcernedly just out of reach while he waited for the hunter to realise that nearness in point of distance is not necessarily the same thing as nearness in point of accomplishment.

In fact the fox's intended victim was so sure the chase was over that

no sooner had he gained his stronghold than he left it again, slipping into the river as quietly as though it were of oil, swimming underwater and with the current, to curve at length into a second tangle of tree roots very like the one he had just quitted. This time, however, he climbed further into the convolutions of the maze as he followed a path worn and polished smooth by constant use.

The entrance to the burrow was well concealed, about mid-way between water level and the top of the bank, and it was musky with the scents of habitation. It was utterly rayless, too, although what to diurnal eyes could only be intense darkness might well have been no more than a pleasant semi-gloom to the homecomer as he followed the curves of the steeply ascending tunnel. There were already a full-grown female and three young lying amongst the dried grass and shreds of bark lining the nesting cavity and here, at least, there was some light, the faintest of faint rays seeping through from a point directly over-head—from an extension of the main passageway into the open air, perhaps.

Ascending no further, the new arrival curled up in the nearest space, disturbing his offspring, who then disturbed one another and, by scuffling and tumbling about for the remainder of the day, con-trived to keep their parents from settling down.

In spite of so many interruptions and awakenings, however, the head of the family had done with sleep by the late afternoon and well before dark was weaving his drowsy way down the steep shaft to the river—a river now dusky with the onset of night but colourful, too, where it caught the day's last quiet gleam.

With scarcely a sound the water rat's slim, furred form glided from the cover of the tree roots into the open, submerged to pass under some floating weed, returned to the surface and again submerged, so that the V-shaped ripple cutting silently through the sunset-tinted reflect-ions would often drift vaguely into nothingness only to reappear after a short interval a little further upstream. Even though the river was as smooth as glass the swimmer was able to remain almost completely in-visible, so well adapted had his kind become, over countless generations, to their aquatic existence. His nostrils being set high and his eyes like-wise, there was no need for him to reveal any other part of himself when travelling on the surface. His ears, too, had been modified, being very small, somewhat rounded, and lying so close against his slim, triangular head as scarcely to project beyond his fur, which was as dense and as waterproof as an otter's.

When the water rat dived the silver bubbles rained upwards from his coat. Continuing his descent into the watery darkness he reached an ancient tree trunk, black and leathery from long immersion. With strong but slow strokes of his partially webbed hind paws he swam around this submerged stump, picking off one by one the largest of the many pond snails clinging to its sides; and when his mouth would hold no more he set off in an ascending slant gradual enough to leave him with his air supply exhausted by the time he surfaced within the safety of the tree roots. A few moments later and he had placed the modest result of his excursion in front of his ever-hungry family.

He made a number of similar journeys, pausing seldom and then only to dispose of a pond snail or two on his own account or to rest awhile on a ledge under the bank, evidently a favourite haunt of his to judge from its worn surface and the broken shells scattered about.

When he did finally change the nature of his foraging he went downstream, swimming just below the surface for much of the way, his sharp eyes scanning the riverbed; and often he would hang suspended in the slow, near-black flood where the only light was from the stars in the cloudless night sky.

So he hunted, as leisurely and as confidently as a kestrel hovering over the plains in bright sunlight. It was an effortless mode of travel, requiring only the occasional sweep of a webbed hind paw to keep him on course, and providing a moving vantage point from which to swoop down on the small freshwater crayfish, or yabbies, much tastier than the pond snails which so far had made up the largest part of his diet. Not that the yabbies were easy to catch. They were as quick as birds and every time he dived his quarry would skim across the river floor and vanish into a hole in the bank, leaving only a puff of silt to disperse gradually in the current.

A crescent moon, added to the glimmer of the stars, did little to help relieve the darkness as the small voyager continued on his way, now ruffling the calm river's surface, now leaving a vertical trail of bubbles as he sank in search of anything worth the taking. Abruptly quitting the water, he foraged for a while amongst leaves and other crackling deadstuff on the bank but finding nothing, was soon swimming towards a reed-bed he had visited before and knew well. A pair of swans were nesting there and though the great sharp-eyed birds were feared by the small hunter it is also true that nests are sometimes left unguarded.

Though placed deep within the reed-bed the nest was also in one sense unconcealed, every growing thing around it having been cleared away to leave the central heap of sedges and interwoven green stuff as conspicuous as a fort surrounded by a moat. And whether the fort and the moat were formed by chance or by design there was no doubt that their existence served the purpose of denying any vestige of cover to an enemy, whether he made his approach by air or water. It was from the edge of the cleared space that the stealthy adventurer finally examined the way ahead and the outline of the nest against the sky. This night, for the first time that season, the silhouette of a brooding swan was missing.

The golden water rat edged cautiously forward—only to find that the seemingly guileless moat was by no means easy to cross. Every one of the reeds the swans had used for building had been shorn off just below the surface; and the submerged stalks left were so tough and close-packed that progress through or over them was of the slowest and most difficult. Swimming was impossible, and the would-be despoiler of the nest was eventually obliged to clamber awkwardly over the sharp, unyielding butts of the reeds, managing in this fashion to splash his way forward at a fair rate, though in a manner ludicrously different from the furtive methods natural to him.

Soon, indeed, the labour of making headway across so many concealed obstacles claimed so much of his attention that he completely failed to see the swans floating motionless against the wall of tall, green stalks. They were on him in an instant. Taken completely by surprise the floundering intruder was at their mercy for a measurable space of time and it was only the furious anger of the great birds that saved him. It seemed incredible that he should remain unscathed after being in the very centre of such a maelstrom of savage wingstrokes, yet he was never squarely hit and at last, dazed but unhurt, he dived. But he was unable to dive deep—no deeper, in fact, than his own length—and there he crouched while the swans beat down blindly upon the stout reed butts which now protected as well as encaged him.

After a while, and heartened no doubt at finding himself still alive, the water rat started to squirm and squeeze between the close-growing stems. He had not gone far, however, before he was forced to come up for air, his reappearance being the signal for a renewal of the onslaught which had gradually begun to subside.

Three times more as he worked his way laboriously towards safety he had to present himself as a target and three times the relentless black flails, tough as whalebone, whipped the water where a moment earlier he had been. It was the crooked first joint of the wing, used as a cudgel, that gave such force to the blows, any one of which, landing on its mark, would certainly have knocked the life out of the small trespasser. When he emerged next he was in a part of the reed-bed where the crowded stems grew strong and straight and high. He could hear the swans still rampaging along the edges of the moat and as soon as he had regained his breath he continued his flight, though at a more deliberate pace, until at last the reeds thinned out, giving him room to move. As swift now as the speckled trout he zig-zagged for the open water and, swimming fast, put yet more distance between himself and the clamour of the swans.

When his sharp, bewhiskered nose, marvellously inconspicuous, thrust into the air again there was comparative quiet. He floated awhile motionless in the black shadow of the bank, then with scarcely a ripple to mark the action, climbed on to a tree root. The business of finding food for himself and his family was proving no easy task this night.

But when at length he abruptly sat up and scratched the base of his skull with his webbed hind foot it was a sign that as far as he was concerned his last adventure had never been. Re-entering the river he headed back against the slow stream, not in defeat but searching as keenly as ever, and was rewarded at last by unexpected success, modest though it was. He was paddling quietly along when, under his very nose, a yabbie emerged from a hole in the bank to be captured with efficiency and promptness.

Without attempting to eat any part of his prize the water rat made for his home burrow; and his home burrow, not surprisingly, was exactly as he had left it—too crowded with boisterous, over-active youngsters, who soon reduced what had been a substantial catch to a few fragments of carapace. Almost as though this ravenous exhibition had forced him to a decision the head of the family and its sole provider immediately led his dependents towards the river.

For the young ones it seemed an extremely casual introduction to a strange and dangerous environment. They did not even pause at the mouth of the burrow but boldly emerged, the whole procession of them, into a maze of twisted tree roots where dark ripples lapped and gurgled in crevices and against the soft earth bank. A few wind eddies, born mysteriously out of the vast calm of the plains, were ruffling the surface of the river and setting the leaves of the red gums rustling so that the night which had been so quiet was now full of small sounds.

Having safely negotiated the well-used path through the tree roots,

47

the water rats swam in line out into the river. Then they returned to the bank in a long curve and landed.

As soon as he had settled his charges on the ledge he had visited earlier that night the leader of the group submerged, leaving his partner the responsibility of keeping watch—a responsibility she seemed to take rather lightly, to judge from her preoccupation with a fur cleansing and combing ritual which took so much time and effort. Paler than her mate as far as her upper surfaces were concerned, she was almost white underneath and, as she twisted and turned in grooming herself, she would seem to become faintly luminous occasionally in the soft gleam of starlight. When her partner climbed on to the ledge she in her turn dived silently into the river. So they shared the tasks of watching and foraging.

Their offspring were a restless lot. When there were pond snails to be shared they scuffled over them and while they waited for one or other of their parents to emerge with a further supply of food they played almost as strenuously as they had fought. With so much activity all around it might have been expected that they would become something less than vigilant, yet the thin, sharp snap of a dry twig stopped them instantly, in mid-movement, and a moment later the ledge was empty. Their going had been quiet and swift, the only sign of their present whereabouts being the dark glinting, some distance away, of a swirl in the water as the three young ones, less careful than their parents, came up for air. A gentle rippling, not unlike that made by a shoal of small fish, continued to move away as the fugitives widened the gap between themselves and the dangerous bank.

On this occasion they were led to a dead sapling stranded in a shallow backwater under the lee of the massive peninsula of flood debris. Being fairly well-grown, the sapling had become anchored in the mud by some of its larger branches; and yet at the same time its butt chanced to be floating over comparatively deep water. It made an ideal resting and diving platform.

Very soon the former routine was again in full swing with, as before, one of the parents always left on guard; and it was as well that this arrangement was not abandoned, for all the movements on the sapling were being broadcast over the sheltered tranquillity of the backwater as a series of tiny, silvery lines—a faintly shimmering web of them—that quickly attracted the attention of a brown and grey owl. Leaving a thicket of low trees on the river bank it drifted into the air and went to investigate more closely this evidence of the unusual.

48

The situation it found was entirely acceptable, with the largest of four water rats about mid-way along the main stem of a sapling, seemingly half-asleep, and three smaller ones, also oblivious of any danger, grouped conveniently together near the butt. There was no question but that here was desirable quarry, and the owl swooped to claim them.

The riposte was a miracle of speed and determination. So swift, so courageous, so unexpected the dash along the sapling and the defender's upward leap that the bird was nearly torn down—as it was the water rat's long, razor-keen front teeth clipped away a wing feather before the owl's alarmed reaction could lift it out of reach.

The feathered hunter's second chance to make a kill came almost immediately—when the young ones in their terrified efforts to huddle even more tightly together suddenly slithered over the curve of the sapling's smooth stem into the water. Then, in scrambling back, they presented three separate and widely-spaced targets; and the owl was quick to descend again, threatening first one then another of the three without ever being quite certain that the opportunity was there to make a capture without risk.

For the protector of his brood was all defiance and agility. He seemed to be everywhere. He looked twice life-size as he parried every thrust and whirled or raced to meet every lunge with the white flash of bared teeth. At last his three charges had regained their positions and were again huddled so closely together that they could not be taken without first overcoming their guardian. So the owl rose out of range—but only to launch a series of mock attacks.

The bird was patient, silent, and almost ghostly in the gentle buoyancy of its flight—even in attack it was soundless—whilst the water rat by contrast was lithe and quick-moving, his claws rasping the sapling's bark as he drove himself mercilessly to meet every onslaught. Once, at the end of a headlong plunge and when it seemed the attempt might well be successful, the illusion of featherdown softness in the bird was suddenly dispelled by the unfolding of a set of powerful talons at full stretch, and ready, longingly, to clutch the life out of the quarry. Then, as suddenly as they appeared, the armoured points were sheathed and drawn up into a mass of thick soft feathers.

Repeatedly, the owl made as though to press home an attack only to sheer off at the last split second as, cleverly and sure-footedly, his antagonist blocked each approach until, at last, the inevitable happened—the bird cut its margin of safety too fine and the fierce little animal's teeth closed solidly on a bunch of wing quills. There came a

49

hissing cry from the owl, a flurry of wide pinions and both challenger and defender splashed into the river.

The water rat dived, circled as expertly as a fish, and speeding back to the surface regained his place on the sapling in a continuation of the same smooth manoeuvre. With an upward point of his head he sent the spray misting from his coat, scarcely bothering to glance in the direction of his enemy now desperately flapping in the river like some great grey-brown moth that has skimmed too low.

In the midst of this small commotion of spray and threshing wings the second of the water rats drew herself on to the floating platform. It would have been an easy matter for them to have made sure that one bird of prey at least did not live to trouble them again but they seemed to have lost interest; and it was simply good fortune that brought the owl, struggling more feebly now, into contact with the dead foliage of the sapling's crest; and it was again by fortune that the spread, exhausted wings chanced to lie at last like pale fingers across a mass of fine twigs strong enough to give support.

Slipping noiselessly into the water the leader of the family group now quitted the scene, the others following, instructed perhaps by some sign imperceptible to alien eyes, for there was no detectable difference between his demeanour on this occasion and when he had left on his many foraging excursions below the surface. Nor did he make any sound. There was no mistaking the eagerness of his suite, however, to press close as he led them towards the shelter of the tree roots and the safety of their burrow.

When, a short while later, they were back in the nesting cavity, he went higher, into a similar chamber, evidently near ground level for its roof was fuzzy with small grass roots. Satisfied that there was no break in the safety system of the home burrow he returned the way he had come, and on down to the tunnel's mouth.

It was almost daylight as he swam towards the tree trunk bridging the narrowest part of the river. Here, he took up his position of the previous morning and once again started to comb his silk-fine coat.

A short distance away the owl had drawn itself out of the water, struggled through the leafy top of the sapling and gained a perch on the thickest part of the floating stem in much the same place as that occupied earlier by the water rat. But the bird, its feathers clotted into tufts and ridges, bedraggled and disconsolate, was now in direct contrast to the sleek and composed little creature grooming himself in the first light of day.

Sometimes the castaway in its efforts to dry its feathers would draw the water rat's attention for a while because of the sound of its wet flutterings; and then, unexpectedly, the bird was flying, barely airborne at the start but, later, as the spray smoked from its desperately-beating wings, gradually gaining height. It managed to clear the bank and to rise yet a little further before it vanished into the trees.

The water rat lost interest again—but only until the same breeze that was bringing the dewy scents of a new day brought also to his sensitive nose the warning taint of fox.

Shining now like burnished metal in the early sun, almost as bright at that moment as the gold in his own name, the golden water rat whisked about and darted smoothly along the dead tree trunk, paused at the point where it reached the other side and with a last resplendent flash, swarmed down the vertical bank to safety. Nor did the fox attempt to follow any further but, having completed his crossing just too late to make a kill, trotted on through the shadows of the red gums. From a drover's camp somewhere along the stock route came the distant barking of a dog and the fox hurried a little.

Brown reflections quivered when at length a small slim shape glided out of the tangle of tree roots to turn lazily downstream. The swimmer submerged before he reached the next scoured-out labyrinth and surfaced only when he was deep within its concealing shadows. Light flickered briefly on his shining, wet fur as he climbed towards his burrow, then there was only the dark.

But outside in the vivid world of day the river, with its glassy stretches, its green reed beds, its massive, convoluted red gums, its depths, its shallows, its grave reflections and its gay, wind-ruffled curves, warmed and sparkled in the ascending sun.

# *Pirate Coast*

SEAWARD the air was clear, even down to the horizon, where the blue of the ocean was as intense as that a mile or so offshore. The inland hills were hazy with smoke that hung like a curtain between the sparkling edge of the water and the hinterland.

There were always fires in the hills—on the plains, too, for the Aborigines moved from camping ground to camping ground, burning off as they had done for ages, thinning out the undergrowth in the forests and flushing the small game on which they lived.

It was not possible to see very far west that day, even from the top of the headland, but the view up and down the coast was dazzling, with grey-green slopes plunging down to the sea, dark cliffs and the brilliance of blue water and blue skies.

The headland itself was of granite. Rising to a spectacular peak it was overlaid in places with enough soil to support a few patches of scrub and some stunted, windswept trees. The forbidding lines of the dominant crag were still unsoftened by any wisp of green; there was nothing but stark granite—and what looked like a tangle of sea wrack or a heap of forest debris. From such a height the world moved slowly. The Pacific rollers breaking in the shallows drifted lazily up the sloping beaches. The murmur of the sea was drowsily monotonous.

Except for a few gulls floating near the shore the surface of the ocean was empty. Neither Aboriginal canoe nor other craft was within sight and no voyager had passed that way since a flotilla had come down from the north. That was a full year ago.

They had landed on a nearby islet to rob the shearwaters' burrows and had put out to sea again with their stores of fresh food replenished. Flocks of birds were always wheeling over an islet, two or three minutes of leisurely flight offshore. With the exception of the gulls

floating near the cliff's base there had been no other living creature
visible until an osprey came down out of the sun in a steep swift slant,
dipped below the level of the headland and then, sweeping up on his
long raked wings, alighted with a flourish on a nest of sticks near the
loftiest part of the crag.

The osprey arched his wings, ducked his head once, twice, then
became still except for slight movements as he brought his gaze to
bear upon some object either floating or flying above him at a great
height. Though considerably larger than most hawks he was smaller
than the blue-grey sea eagle which often went by on its beat up and
down the coast. He would have been about the same size as a smaller
predator, the red-backed sea eagle with the white head and neck,
which also ranged over that territory.

Brown above and white below, with brown mottlings on chest
and lower throat, the osprey was a handsome, vigorous-looking bird.
His flight feathers were black and when his wings were folded their
tips overlapped the end of his tail. The white crown of his head
merged into a short crest of dark feathers that continued along the
hind-neck as a mane, increasing the appearance of alertness. Wide-
spreading claws with curved talons, powerful legs and the hooked
shearing beak of the flesh-eater made up the usual armament of the
bird of prey and added the quality of fierceness to those already
suggested. Only the face and especially the eyes were distinctly
different from the other hawks and eagles. The face was not as narrow
and the eyes had a much wider range of vision because he had not
those prominent overhanging brows which lend such intensity to the
stare of an eagle.

Perhaps eagles do not need the all-round vision of the osprey.
Certainly, it must be seldom that they concern themselves with what
is going on above their heads, but it was there that the osprey concen-
trated his attention. High up, at such a height that they must have
been unidentifiable to almost any other watcher, four black specks
were swinging in wide circles.

Nor did the osprey interest himself in any other quarter until some
gannets, having worked their way along the coast came to search the
waters near the headland. Some of the newcomers were white with
black edging to their wings; others, slightly smaller, were mostly
brown. Following one another at regular intervals they beat through
the still air, planing sometimes as they turned and allowing the line
of their patrol to drift steadily north.

It was when they reached the pale green line running from the headland, and marking the existence of a submarine reef, that the searchers found a school of fish and immediately gave chase. No sooner had the leading gannet banked steeply, steadied and plunged headlong than the whole company was spearing down like so many falling crosses to disappear in a series of small explosive splashes as white and brown fishers alike made the most of their opportunities.

Then the gulls arrived, and the terns, the wideawakes and the tropic birds, until the scene which had been so calm became a whirl of colour and the murmurings of the ocean were lost in the screaming of the excited host. They were the more vociferous, perhaps, because the quarry was swimming too deep for most and only the tropic birds, besides the gannets, were able to catch an occasional fish. The gannets aided by their wonderful swimming ability, made capture after capture. Of course they were unable to remain beneath the surface for long, but their initial plunge carried them so deep and they were so swift in pursuit that they were seldom short of time. It was routine for them to bounce back to the surface, float for a second or two before take-off, gain height and go hurtling down as before. It was only because their heads and necks were protected by a subcutaneous system of air cells that the big seabirds were able to sustain the repeated shocks of impact as they ripped vertically into the water.

If the fish were swimming so deep that the gannets were obliged to hurl themselves down from such a height then it was certain they would be far out of reach of the watching osprey and he merely ruffled his feathers and settled down again in the warmth. Perhaps it was the absolute calm of the mid-summer's day that kept such a restless and energetic hunter at rest for so long, but he gave no sign of wishing to leave his place.

Full fed by this time most of the gannets were making off to rest when the frigate birds swooped. The osprey screamed angrily and raised his crest as the first of the raiders, in an incredible exhibition of a vertical dive, expanded suddenly from a midge-sized blur in the sky to a dark-plumaged bird with the wing span of an eagle. Ordinarily, the gannets were fine energetic flyers but now, burdened with their recent catch, they were absurdly outmatched by the frigate birds.

Flashing across the path of the gannets the leading frigate bird was followed in quick succession by his three companions. The four then began such a display of blinding speed and manoeuvrability that the fugitives seemed to be stationary, moving so slowly through the still

55

air that they might just as well have been hovering. The gannets began
to haul themselves higher on labouring wings, though in fact they
were already at the mercy of their tormentors.

The magnificent wingspan of the frigate birds, their sombre colour,
long forked tails and strong savagely-hooked bills all added to the
intimidating effect of their astounding powers of flight, and which-
ever way the gannets turned there would also be one or more of their
terrible foes to harry them. They crowded in, jostling the fugitives
and pecking at their heads, all the while uttering harsh, scolding cries.
Occasionally, one would charge down head-on, the scarlet patch of
bare skin on its breast flaring like a danger signal, to swerve aside a
split-second before reaching the point of collision.

It could not last and suddenly one of the gannets began spilling
fish. Down they dropped in a silvery line and down went the sea
hawks to snap them up in mid-air.

The assault was resumed, the pecking, wing-slapping, and bullying
until all the gannets had been forced to disgorge and the frigate birds
were so distended with food that some of the booty fell unheeded into
the water.

56

Yet, despite the buffeting they had received, the despoiled gannets had resumed their patrolling of the reef area almost before the plunderers were out of sight and were setting about the work of re-filling themselves with fish as calmly as though nothing had happened. Indeed, so matter-of-fact were they in the acceptance of the act of brigandage which had robbed them of their hard-won catch that the osprey, raising his crest and screaming angrily, had been more indignant. Then he settled down again to enjoy the sun.

Many and varied were the activities which now were taking place in that small corner of the ocean. Tropic birds were nesting on the cliff face beneath the osprey's perch and further round the northern side of the headland was a small colony of gulls. The tropic birds were engaged in courting, nest-building, hatching eggs and rearing young all at the same time. Most of the mated pairs in the group were feeding young but some were still sitting on eggs and at least one pair was in the middle of a courting display at a considerable height above the rock.

The performers seemed to have some special quality of weight-lessness—an illusion probably caused by the dazzling white plumage of two birds floating in a world of brilliant blue and strengthened by a frenzy of wing-waving that had no recognisable relation to flying. Two scarlet quills trailing beyond the tail feathers of the tropic birds were as bright as flashes of fire and the strange beauty of the display was in no way diminished by its accompaniment of piercing and discordant cries. Swooping dizzily towards the water the pair rose again swiftly and smoothly, then disappeared around the cliff's shoulder. There were many rocky ledges in that particular place, all crowded with tropic birds in various stages of the production of young; but there was always room for one more nest.

Stretching his long wings the osprey then brandished them as though at an enemy, and with his head cocked to one side kept his golden yellow eye fixed upon a point in the southern sky. It was from there eventually that another osprey appeared flying low with lazy wing strokes towards the northernmost limit of his territory. There was no exact boundary but one thing was certain—if the alien alighted anywhere on the headland he would most certainly be trespassing.

Apparently the newcomer was prepared to trespass and it seemed, too, that the rightful owner was prepared to condone, provided no approach was made to the immediate vicinity of the nest site. Both birds were males and there was a certain amount of scolding and crest-raising. But soon both settled down coolly enough, the stranger on the horizontal stub of a small dead tree some distance back from the headland's peak, the other in his earlier drowsing attitude.

The sun was now well past its zenith and colours were darkening, gaining in richness. Spray rising from the cliff's base no longer leaped up in sparkles of light but hung in the faintest of mists above the rocks. Only the pale, translucent green over the submerged reef had not changed, though it had become more conspicuous because of the deepening shades of blue on either side.

Out to sea a few white caps were flaring and fading, and a breeze rustled the fringe of dried seaweed around the huge, untidy nest on the headland; the resting osprey bestirred himself sufficiently to look about and to scan the sky again.

Activity over the islet was increasing and the shrill clamour of bird cries, carried by the light wind, became audible. Gannets were quartering the waters near the islet, on their way south again in their unceasing quest for fish. A third osprey, riding an updraught from the sea's surface, rose into view over the crag's rim, poised briefly, and landed on the far side of the accumulation of nesting material. The last comer was a female and, after a half-hearted display of indignation, her presence was accepted. Both intruders had seemingly established themselves as close to the original holder of the territory as was safe, for there is no doubt that the male would not have been allowed to alight on the nest itself unless he could do so by right of conquest.

Clouds of wideawakes from the islet together with a sprinkling of other terns had gathered over a patch of ruffled water and their numbers grew steadily as did the volume of their cries until the noise was deafening. Yet the bewildering whirl of seabirds made a beautiful and animated picture in the slanting rays of the afternoon sun, especially the clamorous wideawakes with their slim black wings and white breasts. Many were swiftly rising and falling as they wheeled through the crowded air, others were hovering just above the waves, picking off the shrimps or tiny fish as they leapt from the water to escape from predators below. Unlike many of the seabirds the wideawakes seldom touched the surface in feeding and generally were able to satisfy their hunger without getting their feathers wet. When it came to flying they were as expert as most.

Also they had enough confidence in their combined strength to shriek a defiant protest at the two ospreys who had come to find out whether the larger fish which were attacking the small fry from below were themselves worthy of attention. A few of the wideawakes even started to harass the ospreys, but the bigger birds were already on their way, the male heading back in the direction of his own territory to the south, the female following the pale green shallows over the reef. Both flew low with deliberate wing strokes alternated with short glides.

Another hour must have passed before the osprey who had been perched on the headland for most of the day suddenly took wing. His early morning expedition had not been successful and he was hungry. As a start he flew over to where a pair of brown gannets were catching fish a short distance on the ocean side of the islet. One of the gannets tried to drive him away but the osprey wafted steadily higher and the gannet rejoined his partner. Cruising at about 100 feet the gannets beat into the wind, turned and came back on almost the same path, over and over again. Every now and then one or both would steady, then hurtle down and vanish. The osprey rose further, watching. There were fish but they were too deep for him. Also, they were small. He could clearly see the gannets as they shot down into the depths, leaving a wake of bubbles, and swam swiftly to snap up their quarry. On returning to the surface they would float for a while before taking off and climbing back to their favourite patrolling level which was considerably above that from which the osprey liked to make his attacks. In the present case the operations of the gannets profited him nothing.

So, all concentration, the most efficient predator of them all went down from observation level to hunting level—as did the high-sailing frigate birds. They were as well acquainted with the behaviour of ospreys as they were with that of gannets, cormorants or terns. Stroke after leisurely stroke the osprey travelled over the sea a scant thirty feet above its surface. He sighted one school of fish but it also was swimming too deep. Still he waited, hovering. Above him the frigate birds circled on motionless, drifting wings. Although they had descended from their former altitude they were still at a tremendous height.

The shoal of fish sank until it was no more than a shadow even to the eyes of the bird of prey. Then, as though menaced from below, it rose towards the surface. The osprey balanced in the wind, and plunged—talons extended, fierce head thrust forward, wings hooded. Disappearing in a burst of spray the osprey soused clean under. In a few moments he reappeared on the surface, his wings threshing.

59

Shaking the water from his plumage he rose with a fish in his claws. Unlike the gannets who pursued their quarry by swimming the osprey had to rely on the accuracy of a single strike. It was the measure of his skill that he rarely missed. But the catch, in this instance, was heavy— too heavy to lift up to the headland, even though the bird's talons had already put an end to the victim's struggles.

Carrying the fish head foremost, the osprey made for the beach on the east side of the islet and was fighting to keep above the surf when a frigate bird flashed past. The second raider was travelling faster and came closer, clipping the osprey with its stiff wing feathers. The two remaining attackers whizzed within inches of their slow-moving target. The four then gave such an intimidating display of aerial hustling and bustling that the osprey wavered and landed sprawlingly on the sand. The fish slithered away; the frigate birds descended in a cluster and tried to snatch it up but it was lying too close and was too big to be lifted easily.

Magnificent in flight, swift enough to catch a falling fish when it seemed no power on earth could prevent it from splashing into the sea, expert even in picking small fry from the peak of a wave, the frigate birds were nonetheless wary of being forced to alight, for their legs were not only small and weak but set so far back that walking was impossible. Despite their magnificent wingspread—in another sense, because of it—they were quite unable to rise into the air from the ground.

So their former boldness now gave way to what might have been described in a smaller bird as a nervous fluttering, but which in the case of the sea hawks was more like the threshings of huge black bats tethered to earth by invisible cords. Besides this, the osprey had recovered from the initial shock of the assault and pounced determinedly upon the fish. He stood over it in defiance, claiming it with one powerful claw.

Wideawakes from the islet began to swarm excitedly over the disputants and were joined every moment by other seabirds—noddies, tropic birds, gulls and crested terns—whose cries were so loud and shrill that the harsh croakings of the frigate birds and the strident screaming of the furious osprey were overwhelmed.

Lifting the heavy fish as though its weight were of no consequence the osprey reached a height of about fifty feet in the first strong surge of power. His intention was to fly over the forested crown of the islet and make for the headland, but the densely-packed thickets of pisonia

trees together with the myriads of wideawakes that seemed to fill every available chink of air space proved obstacles beyond his power. Certainly he lost headway, to such an extent that the frigate birds swooped again, cutting through the wideawakes, one of which was dashed lifeless to the ground by the bone-hard wing of an attacker.

Suddenly, the osprey had had enough. The fish dropped, balanced a moment on a bunch of pisonia leaves then slid off. The two frigate birds dived together but balked one another; the osprey made a belated attempt to recover the booty, missed and slashed at the frigate birds with his talons. There was a three-cornered collision and the osprey and one of the frigate birds found themselves on the floor of a

61

shadowy pisonia thicket. Standing tall on his rangy, widely-set legs the osprey half-sprang, half-flew to a branch not far above his head and launched directly into the air to rise high into the late sunlight. Below in the gloom of the pisonia thicket, the frigate bird was crouched on the ground. The nature of its plight became evident when it started to move. The sea hawk was almost helpless; because it spent nearly all its life in the air the frigate bird's legs and feet were practically useless and its only mode of progress on the ground was to lift up the forepart of its body by using its bent wings as crutches then to push with its tiny feet until it fell heavily forward on to its breast. Thus, slowly and painfully, it made its way between the trees, rising and tumbling in a bizarre crawl that brought it eventually to the edge of the thicket and the top of a sloping, sandy beach.

For a while the great bird lay motionless, before trying to take off. Its first two attempts to rise ended in failure, but the third time a single, shortened wing-stroke was successfully completed without making contact with the sand and, genius of flight that it was, the frigate bird lifted clear and a moment later soared into the sky as though hurled aloft by a whirlwind.

Late afternoon was merging into dusk when the osprey, having traversed a good deal of ocean since the encounter with the frigate birds and having made no other kill, rose from the level at which he had been flying and set course for his accustomed nesting site on the headland. Swift, effortless and graceful in flight on some occasions he was now beating heavily homeward with the phlegmatic, flapping wing beats of a raven. There was at least one osprey in that locality who would sleep hungry that night. Sweeping up in a current of air streaming over the cliff's edge, he came to rest with outspread wings, remaining so for a moment or two as though striking an attitude.

Within a short time the same female osprey who had visited the nest earlier in the day returned and alighted in the same place as before. A partly-eaten fish was gripped in one claw. Nor did she show any inclination to resume her meal but merely settled down with one foot set upon the prey. She had made a number of kills that day and was full-fed. She quickly roused from her lethargy, however, when the male osprey approached, bobbing his head and uttering a series of cackling cries obviously intended to be ingratiating.

The other gave no sign either of pleasure or displeasure but as soon as the suppliant had sidled to within range she flicked out a wing and dealt him a blow that left no doubt as to her reaction. The owner of the nesting site did not press his claim. As with most of the hawks and eagles the female is bigger and stronger than the male and if this particular female would not be coaxed into giving up the booty then that would be the end of the matter.

He stayed close to her, all the same, in the hope that some opportunity of taking the edge off his hunger might present itself. But it did not. She tore at the fish with her formidable hooked beak but ate nothing and when he again reached tentatively towards the remains of the kill she struck at him as before and, advancing angrily upon him, allowed the fish to fall over the edge of the nest where it lay on the rock unnoticed.

Now that the cause of their bickerings had gone the two ospreys became indifferent to the presence of one another. They stayed there, growing more shadowy as darkness fell, one on each end of the nest pile. They stood there as still at statues when the stars came out to relieve the intense blackness of the sub-tropical night. The sound of the waves seemed to deepen but the crying of seabirds on the islet were as loud and as strident as ever. Sometimes a flock of wideawakes would wheel over the headland and then their cries would be deafening when no longer tempered by distance. But the ospreys never stirred, not all through the changeable night which was in turn black then beautiful with moonlight, then dark again with storm clouds and starlit for a brief space before dawn.

When the moon rose the ocean was silver and the islet of birds as black as coal. When the storm built up behind the inland hills and blotted out half the sky, when the thunder crashed and the lightning stabbed at the earth it was for a while also moonlit in the east. But shortly the moon was overwhelmed and the crying of the seabirds grew noisier.

There was nothing gradual about the rain. It came down with a roar that could be heard in the distance, quickly enveloping the headland and spreading out to sea. When the lightning flashed the rain was like falling spears. Water spilled in rivulets from the osprey's nest and streamed in torrents from the cliff. Yet the wideawakes were still flying and calling. The glare of lightning showed the rain swirling into space from the glistening wet rock ledges and the tropic birds sitting impassively on their nests. The two ospreys had hunched their shoulders to the downpour, but otherwise remained unmoved. The rain trickled from the glossy sheath of feathers which was nearly as water-resistant as that of a gannet.

Dissipating as quickly as it had gathered, the storm was over long before there was any light in the eastern sky. Seabirds are never slow to waken even when some of them have been flying and calling through the night, and by dawn a flock of seagulls was squabbling over the flotsam that had been washed on to a nearby beach, the tropic birds were floating wraithlike in a wan, grey world, their discordant cries adding to the noisiness of the gulls and wideawakes and the squeals of the crested terns.

The female osprey was the first of the two to set out again on the unending quest for food. Spreading her long wings she sprang into the air and sailed over the abyss, to fall lazily towards the water gaining speed and finally rising in an elegant, rising arc that swung her high above the crag where she balanced in the light wind, hovering with short, beating strokes. She gave a whistling call then keeled over to go planing down towards the waves, her brown plumage showing in sharp relief against the sparkling water now touched by the sun.

In spite of the fact that he could not have fed for the past thirty-six hours or more the male osprey seemed more interested in his environment than in beginning the day's foraging. His golden-yellow eyes were fixed on the islet and never at any time seemed to concern themselves with the myriads of seabirds surrounding it. But he followed every movement of the four frigate birds that rose in succession from roosts in the pisonia thicket and cut recklessly through hosts of lesser creatures. They were more like strange prehistoric animals of millions of years ago than developed creatures of today. Their sable plumage had a leathery look, their long, strong bills and small heads were not unlike the bill and head of a pterodactyl and their harsh croakings were different from the shrill, squealing calls of so many seabirds.

As they circled higher they attracted a deal of notice from the smaller birds and a few attacks were made upon them. Plunderers and despoilers they were but their very speed and flying ability made them dangerous opponents and none of the protesters ventured too close.

Emerging from the press the frigate birds climbed grandly, drifting apart as they rose. They shrank from the size of eagles to the size of hawks, of skylarks, of dragonflies, of midges. To ordinary vision they were then swallowed up in the blue, but the osprey continued to watch.

While the sun was still low in the east the osprey also began his daily quest for food, flying low, his wings flapping for a spell then stretching out in a slow glide that cost a minimum of effort.

His first essay was a crashing dive that sent up a spout of white water. Widely-spread talons covered the moving target as the hunter plunged deep, his fierce head thrust so far forward that his beak was almost as close to the quarry as his claws. Again the prey was big, bigger than the kill of the previous day, and for a while it was doubtful whether the fish would be lifted to the surface or the osprey dragged to the bottom. The sharpness of the bird's talons was the deciding factor, that and the grasp that drove them in with so much power. The captive's buckings ceased and the osprey's head and shoulders emerged, supported by two weakly-flapping wings.

But he was by no means spent and, after an interval of quiescence, the long pinions lifted full-stretch then beat their way upwards. Not to remain in the air, however, with such a load to sustain. So the fish was skip-hauled from wave top to wave top and in this way was landed finally on the beach next to the headland.

The usual swarm of smaller seabirds gathered above the osprey's head but he ignored them. There were no frigate birds. For once they had not noticed a happening that could have been turned to their advantage. The osprey started to feed, holding his capture under the claws of one foot and tearing off piece after piece with his hooked beak. When he had eaten about half of the fish he flew into a dead tree with the remains, only to allow them to fall into the scrub growing beneath.

Summer blended into the milder days of autumn which, in a land where one season is not vastly different from another, meant only that the warm evenings were hazy with mist and not with heat. The number of ospreys roosting every night on and about the headland had increased to about ten and this loosely-knit colony lived amicably together for the most part, chivvying one another occasionally, as some flicker of territorial rights passed through their avian brains but never pressing a claim or an imagined claim to the point of serious disputation.

It was only when a sea eagle came sailing into the neighbourhood that the ospreys showed unmistakable signs of hostility. Not by any standards could the sea eagles be regarded as skilled fishers. They were not to be compared with the ospreys, either for industry or accuracy or determination in diving, but they *were* capable robbers— once again not in the first rank with the frigate birds but not far behind. On occasions too, the ospreys were able to give the sea eagles similar treatment to that which they often suffered. The main prerequisite for success was that there should be a party of ospreys and only one eagle; it was also necessary that the eagle be heavily laden because if the prey was small he could soar above them.

It was in autumn, too, that the eggs of the green turtles were hatching on the beaches, with thousands of tiny replicas of their ponderous parents running wildly towards the waves. Very few reached their destination for the advent of young turtles was a feast day for predators—terns, gulls, ospreys, eagles, frigate birds—who snapped up most of the helpless scurrying creatures before they came within reach of the equally hungry dwellers in the sea.

Life resumed its usual pattern after the bonanza of the turtles, the ospreys catching fish and being robbed of their prey as systematically as ever by the frigate birds, and to a lesser extent by sea eagles. The frigate birds menaced every bird in possession of a fish, whether they were large or small or whether they were carrying their catch in their beaks, claws or bellies. They bullied the tropic birds, often struck down a tern or a gull, harassed the gannets until they disgorged, badgered the sea eagles and never failed to attack the ospreys.

The ospreys suffered most. They were such efficient finders of game and so successful in its capture that they were the most favoured recipients of the attention of every avian pirate in the area. Though many of the seabirds were highly specialised fishers, each in its own way, the proportion of misses made by every one of them except the ospreys was so great that for much of the time only the ospreys—and, possibly the gannets—were worth persecuting.

In fishing, the ospreys were without peers. Whether the situation called for the delicately judged snatching up of a fish from out of the deep trough between waves or whether it demanded a violent crash dive, they seldom missed. Therefore, whenever a high-sailing frigate bird saw an osprey touch blue water, either gently or in a burst of spray, it would hurl itself down in the near-certain knowledge that the hunter had made another kill.

When the frigate birds began their courtship displays in preparation for the breeding season the persecution of the ospreys became less intense. Certainly, they lost a few fish to the sea eagles whenever these great birds chanced to be in the vicinity and they suffered from occasional mobbings by the terns; but for the most part, they were able to live peacefully.

66

Each of the ospreys which had selected the headland as a base had an acknowledged roost either on the headland or in the hills near the base of the headland, the small colony extending over a square mile or so.

There was only one nest on the headland itself and that was on the rocky crag where the male and female ospreys perched every night, one each side of the mass of sticks and seaweed and flotsam gathered over the years. The nests of all the other ospreys in the colony were in dead trees in the hills up to ten or twelve miles along the coast.

A frigate bird was sailing slowly past the headland, the scarlet patch of skin on his throat inflated to a roseate balloon. Above the islet other frigate birds were displaying in the clear autumn sunshine, the red balloons making sharp points of colour in an animated scene where everything except the blue sky seemed to be moving—the black sea hawks flaunting their red signs, the white gulls and gannets, the black and white terns, the waves and the white caps, the clouds and the green, wind-blown pisonia trees on the islet.

Before the frigate birds had reached the stage of building their nests in the branches of the pisonia trees a change came over the behaviour of the ospreys. The cock birds, and there were six of them, began to show hostility towards one another and after many aerial duels four of the males quitted the headland for the nesting sites they had used in previous years. All the females except one also disappeared, leaving only the male who claimed the nest on the headland as his own, the male who had taken up residence in the stunted tree near the nest, and the female who had been the first of the colony to return and who, in fact, had been mating for some years with the present holder of the territory.

Courtship was a noisy, vigorous and flamboyant affair. It started with a clash between the two remaining cock birds, the tree-dwelling osprey getting the worst of it and withdrawing his claim by changing his nightly roost to a tree at the base of the headland, where he was allowed to remain.

For some days afterwards the victorious osprey gave some astounding exhibitions of flying—spiralling to dizzy heights, to plunge downwards again, level out and hurtle down again towards the female at the nesting site. He demonstrated a sequence which started with a number of shallow dives alternated with short, level flights and ended with a distress act in which the male osprey, then a short distance beyond the cliff's edge, suddenly assumed an erect position

67

in the air, uttering shrill screams as though in the last extremity of terror. Wings flailing wildly and all semblance of control seemingly lost he started to sink faster and faster in a helpless swirl of feathers until a moment before striking the water he stretched out his pinions and glided smoothly across the wave tops to rise again towards the summit of the cliff.

These solo displays later gave place to other exhibitions in which both birds took part—mad chasings, whirlings and tumblings at speed from a great height.

After mating on the exposed platform of the old nest the pair of ospreys set about adding fresh material to the pile that had resisted for eight or nine years every storm that had threatened to wrench it from its foundations. They flew back and forth for nearly two weeks with trailing clusters of dried seaweed gathered from the beaches, with pieces of driftwood, with the desiccated remains of seabirds which had been bleached by the sun, with small dead branches snapped off the high trees round about and any other object that took their eye. The wings of seabirds, provided they were well-feathered, seemed to be the most sought-after material, with dead branches and huge, trailing bunches of seaweed next in demand. Their method of obtaining the branches was to alight heavily on the selected piece in order to snap it off. Generally, they were successful but sometimes the branch would not break, even after several pouncing descents and sometimes, on being broken off, it would prove too heavy to be carried away. When this happened neither of the ospreys let go until the last possible moment and never failed to fight with the branch as with an enemy, displaying the raised crest of battle, uttering piercing screams of fury and even attacking the insensate wood with their hooked bills.

Both ospreys were in the excited state peculiar to the breeding season—pugnacious, suspicious, jealous of their territorial rights. They repelled the ravens when they came too close to the nest, and the male osprey took every chance of assailing his counterpart, the unattached male, who kept edging closer until he was chased away again, yet was never once treated with hostility by the female of the mated pair. A possible danger to harmony might have arisen when the unattached osprey helped the other two drive away a sea eagle, but the paired male promptly corrected the position by turning upon his helper as soon as the eagle had been defeated. Locked together by their claws, he and his rival went whirling down the sky, separating only at a dangerously low level. Pursuit was unrelenting, too, and the victor did not turn back until he glimpsed the holder of the adjoining territory winging swiftly towards the scene in order to continue the assault upon the trespasser.

In the days before and immediately after the laying of three osprey eggs the environs of the nest were lively with constant activity. The noddies and many wideawakes had gone, but the tropic birds, having reared their young, were spending more time sailing colourfully against the blue. The white-plumaged gannets had flown south to nest but the brown gannets were working harder than ever in order to feed their partners now brooding on a sandy strip on the barren side of the islet. Unfortunately for the brown gannets most of the frigate birds had finished their courting displays and were again on the lookout for food for their families. With the ospreys not greatly concerned at the time about catching any fish the brown gannets were being relentlessly persecuted by both the frigate birds and the tropic birds. Second only to the more practised pirates in the power and speed of their flying these sheeny white birds with their burnt-orange coloured bills and their long scarlet tail quills seemed to set the air afire with the velocity of their dives and near vertical ascents.

The first time the female osprey was relieved by her partner from a spell of sitting on the eggs she quickly caught a fish on her own account and was immediately attacked by a swarm of hungry raiders. But the fish was small and she reached the headland safely and managed to eat most of it in spite of having to defend it and herself continually. It was filched at last by a tropic bird while the osprey was engaged with other assailants. Dropping down to the water she ducked her head repeatedly into the waves and dabbled her claws to rid them of the scales and blood of the kill, then sailed back to resume her brooding.

The islet that was such a short distance from the headland was covered with nests, eggs, and young in various stages of growth and it was the prospect of easy pickings that brought the ravens every day to the big rookery. Seldom, too, were these visits unrewarded for the breeding grounds were so crowded that there was always a fight going on somewhere, often two or three, and it was then that the black opportunists wheeled down to the feast, slipping quietly away when they were satisfied and always leaving behind a few nests littered with broken shells or empty of nestlings. Because they were land birds the ravens liked to approach the islet from the point of the headland, thus avoiding a long flight across water, and inevitably this route brought them close to the ospreys' nest. Consequently there were frequent disputes between nestholders and passers-by with a good deal of noise and agitated manoeuvering but not much actual fighting.

Besides warning off the ravens, threatening the sea eagles and any other bird that flew too close to his nesting area the male osprey now had the task of catching fish for himself and his partner. This in itself was not onerous because the female bird ate little while brooding and because the frigate birds were now so hard put to it to provide for young which had appeared in their own nests that they could not afford to hover for hours between sea and sky waiting for the osprey to make a catch. They were obliged to do some catching themselves, if only of small fry. Because of this the male osprey was often successful in getting home with his first catch of the day, which meant that he need not go on patrol again until noon or even mid-afternoon. For ideal conditions of food procurement the frigate birds depended on keeping the ospreys hungry and, therefore, hunting. Permit any one of them to devour his catch and he would fly to some high place in the sun and drowse there until his appetite returned. Preoccupied as they were with the insistent demands of their offspring the frigate birds too often allowed their main providers to become inoperative.

The disturbance started almost without cause. Once again the male osprey had killed and fed, incidentally confounding his natural enemies, the sea hawks, because he had also killed the previous evening and so had not set out on his usual early expedition. When he eventually sailed away from the cliff he was successful in his first attempt and after eating as much as he wanted, carried the remains of the fish up to the nest.

He set about tearing pieces from his catch and proffering them to the female, both birds making soft throaty noises. She ducked her head and quivered her wings. Then she stood up and stretched her long powerful legs, carefully moving away from the eggs, now in their second week of incubation.

Suddenly, at a drawn-out croaking cry and the well known rip of cloven air, the crests of both ospreys rose like the hackles of wild dogs. But the frigate bird was already past. Flying in the opposite direction was a company of six or seven ravens.

Having seen the frigate bird swoop and noticing the osprey eggs uncovered the ravens circled, two of their number diving at the nest, only to be met and chased away by the female osprey. Then the

70

solitary frigate bird returned and also dived at the eggs. This time it was the male osprey who rose in defence and turning on his back claimed the frigate bird with his deadly talons, the thump of impact distinctly audible. It was an unusually savage and daring stroke for an osprey and the two fell headlong, breaking apart just above the water, the osprey to plane smoothly across the waves, the frigate bird to finish in the sea, apparently dead.

The osprey's mate had dived down alongside the combatants and now both birds of prey sailed back to the peak of the headland. From start to finish the engagement had been a matter of only a minute or so. As the two ospreys appeared over the rim of the cliff the ravens fled, but they had had time to do their work and there was only a mess of broken shells left in the ospreys' nest.

For a while the female tried to settle over the fragments but gave up and took up her stance nearby. Continuing in a state of bewilderment the ospreys now combined many stages of the breeding season so far. The male caught fish and presented them to the female, he also tried to feed her with pieces torn from his latest kill and the two of them resumed in a desultory fashion the former work of nest building.

When three more eggs were eventually laid, the behaviour of the mated pair followed much the same pattern as before, although the male seemed even quicker to drive off intruders.

Each of the second set of osprey eggs hatched. Unlike the young of most other hawks the nestlings were not naked but were covered with down which, although short, was fairly dense, buff in colour with a pinkish tinge. All of the three were too weak to lift their heads and the last to hatch was smaller than the others.

The male osprey excelled himself in the catching of fish. He had no choice since the frigate birds had resumed their former tactics of wheeling high in the air, waiting. They seldom failed to rob him of the first three or four kills of each day. The hen bird was as assiduous in her care of the young as her partner was in supplying her with food. The nestlings suffered greatly on hot, sunny days although the female would stand over them for long periods with outstretched wings while they panted for breath, moisture dripping from their open bills. Not long after the quills of incipient feathers had appeared under their down the youngest died, having been pushed aside too often by his stronger brother and sister at feeding times. Weakened by lack of nourishment he failed to survive an especially hot day.

The male osprey started to relieve the female from her maternal

71

duties more frequently than before but her absences from the nest were always brief—a slow glide down from the crag, a vigorous splashing in the water when she would often immerse herself entirely for a few seconds, a shake of her plumage as she took off, a perfunctory quest for game in the close vicinity and then she would be flapping slowly back towards the nest.

About half-way between the time of hatching and the time of leaving the nest—just when it seemed that the young would surely grow into creatures not unlike their well-favoured parents—the coast was mauled by a succession of gales. The bad weather started during the night. The stars vanished very quickly and a storm out to sea lifted its rumbling tones above the voice of the ocean. Lightning showed the rising waves and the blackness of the clouds. Then the rain swept in, drenching the ospreys and their nest. Three or four times the whole scene leaped into shuddering brilliance, all motion frozen as in a photograph—wave tops dissolving into spume, white water clawing at the cliff, then darkness again. The rest of the night was if anything more violent than the storm which had ushered in the rough weather. The darkness was intense. The noise, too, was so great that its components could not be isolated, the roar of surf on the beach, the crashing of waves against the cliffs, the howling of the wind, the deeper drumming of the sharp-slanting rain merging into one continuous explosion of sound.

Dawn showed an ocean grey with rain and lashed by wind squalls whose course as they swept through the chaos of foam and spray could be followed from just such a vantage point as the lofty headland.

There were few terns, gulls or other of the smaller birds to be seen and many of them—the gulls, unquestionably—had sought shelter inland. But the gannets were fishing, working steadily north into the gale. Long, slim, streamlined wings thrusting them smoothly and powerfully into the blast they would beat their way to the limit of their patrol and when they banked they were beautiful to watch, so swift and graceful was their return to the next starting spot. Sometimes they would hang, balancing in the gusts, sometimes they would spear down and vanish without trace for no white splash showed in that boiling sea.

But fish were scarce and their plunges were infrequent. The prospect for other seabirds was much worse than it was for the gannets. If the gannets dived only infrequently then ospreys and others would have to work hard indeed to keep themselves alive. The pirating of fish caught by better fishers was probably the only way to obtain food in such wild weather but the ospreys adopted this method only when the booty was in view, which certainly never happened with the gannets.

If the fish had gone down too deep for ospreys to reach then conditions for surface feeders like the frigate birds were quite impossible. Their best hope was the gannets and high up, the sea hawks were already waiting.

When the male osprey rose into the storm the wind was at its strongest but his skill was so nearly perfect that he went straight up on motionless wings and stayed there, his head turning this way and that, as though he were considering which direction to take. He swung north following the path the gannets had taken but travelling swiftly in a descending line until he was close to the water. Mostly the depths were obscured by foam but he flew on, now using the deliberate stroke that slowed his progress almost to a stop.

Descending in demonic swoops the frigate birds came to remind him of their claims then vanished to leave the bird of prey to pursue his way unmolested. Buffeted, drenched in rain and spray, baffled by the broken, boiling surface the osprey was back at the headland in a comparatively short time. He brought nothing.

Nor in the afternoon was he able to take anything except a small dead fish picked out of the surf. When he bent his head to eat whilst flying back to the nest he found that the fish was starting to decompose so he let it fall and descended again to continue the search. Because of his brown colour he was most inconspicuous when flying low, especially in dull weather. Perhaps it was because of the dark upper surfaces of his plumage that the frigate birds did not come near him that afternoon, though, keen sighted as they were, it was probably more reasonable to assume that they knew from experience that even an osprey was not worth watching in such a furious windstorm. They were much more likely to be harrying the gannets, wherever those dependable fish-killing machines were operating.

The storm raged for full three days, beginning to abate on the morning of the fourth day and during that time the ospreys caught only one fish between them. Yet they ranged far and wide along the coast both to its north and the south. The female visited an estuary some hours flight to the south but was driven away by the sea eagles that seemed to have congregated there in large numbers. She persevered for a while but they pursued her in relays from the surface of the river to a great height and down to water level again, screaming fiercely and repeatedly trying to strike with their talons. Her narrow wings, swept back at a sharp angle from the wrist joint, which is the joint about mid-way along the length of the wing, proved as swift as their broad ones and more evasive. She left, at last, to take up her own most unpromising territory, the weight of her disappointment seeming to slow the heavy yet unwearying beat of her pinions.

Except for the urgency of obtaining food the female would not yet have left her young unguarded. Certainly, they were big enough to be safe from the smaller predators but would have proved easy game for the largest bird other than the frigate birds, the smoke-blue and white sea eagle. Even the ever-dangerous frigate birds might have killed and eaten them provided they had support and enough time to complete their work undisturbed. Both young, however, instinctively kept still and because of this the chance of any passing bird being attracted to the nest was pretty remote.

It was the male osprey who brought in the only food of the period by catching the one fish he had glimpsed since the storm started. Pursued by larger hunters the fish left the water a short way in front of the labouring osprey. Immediately, the bird crash dived. It was an extraordinary manoeuvre to put into action on the spur of the moment. The osprey's claws hit the trough of the wave at full stretch, touched something and clenched. The fish was big but the needle like points must have pierced deep for it ceased to struggle almost at once.

Threshing impatiently out of the water the osprey set out for shore but, weakened perhaps by lack of food, was unable to lift entirely clear so that the tail of the fish kept hitting the wave crests, threatening to bring the flyer down and causing him to falter repeatedly.

The osprey landed clumsily on dunes behind the beach and well beyond the waves which were running further than usual up the shelving sands. With his plumage blown into ragged tufts because he was not facing into the blast he shuffled around to present his front once more, and again became a streamlined bird of prey. The coarse grass of the dunes scintillated with rain though it was laid almost flat by the force of the wind. Behind the dunes a crescent boundary line of bushes—an unkempt hedge clipped and compacted by the weather of years—shook in the gusts and it was against this sombre background that the osprey now stood. He must have been well nigh invisible from the air and yet a frigate bird had seen him and came to balance overhead on black wings that never moved except for small, scarcely perceptible adjustments to the wind gusts. Although the flyer did not alter its position either forward or back, or to either side, it ascended straight up for about fifty feet then came down again so that its croaking cries were distinctly audible to the osprey, now shrilling his defiance. It was a primeval scene, the rain again sweeping down in a heavy shower, the lashing of matted scrub in the background, the roaring of surf on the beach, the spray smoking from the wave tops, the clamour and the violence of the elements, the two predators, savage with hunger, crying their enmity for one another, the bloodied prey still writhing on the sand.

Had the osprey been in the air with his heavy burden the frigate bird would have attacked instinctively knowing the other to be at a disadvantage with both claws holding the kill and both wings too strenuously occupied to strike at an enemy; but with the grounded osprey ready to slash back at him the situation was vastly altered.

Suddenly, the frigate bird towered into the dimness above, almost vanishing and a moment later, the female osprey swooped past with the wind, banked and hung awhile before dropping down beside her partner. Hidden beneath the overhang of mist they both ate some of the fish. The male bird then neatly nipped through the backbone behind the head and rose like a big black gull to be followed in the same way by the female.

Next day the force of the wind abated, the skies cleared and by early afternoon the bad weather had passed; it was then that the ospreys for some distance around again changed the pattern of their behaviour and the headland became once more the centre of the community. But none of the newcomers approached close to the nest. Some of the ospreys were perched on crags near the headland's base and others on dead trees in the forests above the beach. There were also a few ospreys of that season's hatching and they seemed ready to roost anywhere—except near the point of the headland. Whenever they approached the nest they were driven away.

The two young occupants of the nest were nearly ready to fly. Black primary feathers were well grown in their wings, they were approaching in height if not in size the appearance of maturity, and their plumage in general was also comparable with that of full-grown birds, though they were paler in colour and rather more obviously streaked. They were now strong enough to stand for most of the day and showed the unmistakable leggy build as well as the pointed, death-dealing talons of the bird of prey. Still, they continued to crouch

75

in the nest as soon as darkness fell, their two parents standing guard, one each side.

Every day, after they had torn to pieces and eaten the first kill of the morning—and it was nearly always laid in front of them by the male and only occasionally by the female—they would start to flap their wings vigorously, lifting their body weight until their claws were reaching down anxiously for a grip on the nest. Or they would bounce up and down in great, slow jumps with their wings flailing, in what appeared to be an uncontrollable display of awkwardness, but which never resulted in any collision until the elder of the two rose to ten or more times his own height and, on being drifted beyond the edge of the nest by the easterly breeze, panicked, and side-slipping back with a certain amount of unconscious skill, crashed on top of his smaller companion. At the conclusion of each morning's flurry of activity they would doze until the next catch was brought and dropped beside them.

As the size of the young ones increased so did the parents become less protective, neither did they display nearly as much hostility to the approach of other ospreys, provided none took the extreme liberty of alighting on the nest itself. Other varieties of birds, however, were driven away with determination and a show of fierceness. The older birds now absented themselves from the area for longer periods and roamed more freely, both up and down the coast. Usually, the male would return to the young with the first kill he made in the day and he nearly always paid a second visit before nightfall, but the female sometimes flew so far or drowsed for so long on some sunlit perch that it was often nearly dark when she got back. Even when she did not bring food she was never hungry. At any rate she never showed the slightest interest in the partly-eaten fish lying about the nest, even those that had been freshly taken.

The elder of the young ospreys was the bolder of the pair and showed more aggressiveness. On one of his hovering flights over the nest he was suddenly swirled away in an unexpected gust of wind. Instantly he sent out loud, whistling calls of distress. He essayed a landing in the small dead tree but was unable to keep his balance on the branch and, after a frenzied exhibition of flapping and tail fanning and desperate wobbling back and forth, lost his grip altogether and was swept over the abyss.

Within moments his terrified alarm cries had brought one of his parents, the male, who hovered alongside on steady wings, uttering undertones of encouragement. But the youngster was too unskilled, his plight too bewildering, the great void extending all around him too frightening to be overcome in a few moments, and he fell—not headlong but like a leaf, flutteringly, this way and then that. Fortunately, the easterly was strong enough to blow him over the line of white water and half-way up the beach. Just before reaching the sand the raw aeronaut's control improved, probably because of a return of confidence at seeing the end of the journey so close, and the actual landing was carried out in a calm, unhurried and reasonably efficient manner.

Safely down was a good thing as far as it went but the facts remained that the young one was exhausted, that darkness was not far away, that to remain on the ground was to be at the mercy of any four-footed predator that might be on the prowl between dusk and dawn and, lastly, that the nest on the headland might just as well have been in the stratosphere for all the chance there was of his regaining it that night.

The older osprey also alighted, his upright stance and vigilant, yellow stare lending poignancy to the dejected state of the youngling now drooping forward, propped on the joints of his partly-spread wings.

Tensely on guard the mature bird kept watch upon the shadowy bush behind the dunes and the sky, from which the last red tint of sunset was quickly fading. His shrill, whistling cry was both a signal of recognition and a signal of urgency; and his range of vision was such that minutes passed before the female came down faster than a falling stone, yet alighted gently. The two made the same throaty sounds, quite low, that they had made when changing shifts during the period of hatching but in spite of this display seemed to have no way of rousing their thoroughly apathetic offspring.

Dusk was turning into darkness when the young osprey had recovered sufficiently to rise into the air again. For a while he gained height swiftly, then he began to waver and was blown further away from the nest and over a forest of gums. Wings earnestly flapping he flew in small circles, looking more like a big flying fox than a bird, then settled on the lofty branch of a dead tree. Firmly at rest, his own weight keeping his claws clamped tight around the branch, he was safely established for the night. His two escorts were soon out of sight. They dropped out of the darkness on to the nest rim as silently as owls and without disturbing the remaining young one, already hunched up and roosting for the night.

Early next morning the female osprey dropped a fish beside the young one in the nest. Then she flew off into the sunny haze of the coast. Her partner did not make a catch for an hour or more. When he did, however, he flew past the young osprey in the tree to alight on the sand dunes. After some preliminary flexing of muscles the other flapped clear of the dangerous branches and came planing down fast towards the food, landing awkwardly and blundering into his provider and guardian. There was a brisk interchange of blows before the parent stood aside, his crest still raised, his wings still held in the ready position as though on the point of dealing out more punishment.

Inexpert flyer though he might be the tyro had had some practice at tearing a fish to pieces and the way he used his powerful, shearing beak suggested the skills of flying and hunting would soon be his also. However, he showed no anxiety to leave the beach and it was not until later in the afternoon, after soft whistlings and calls from both parents, that he decided to try his wings again.

As before he rose quickly and a casual glance then would have detected no differences between the three birds flying so close together towards the headland. Then, as before, he tired suddenly and seemed to become frightened at the height the three had already attained. What had been a fairly smooth operation quickly degenerated into a wobbling descent by the pupil and an agitated wheeling and swooping by the parents. Nor was the rate of fall slowed and, finally, the three arrived back on the sand.

Yet, within two more days, the young osprey had not only regained the nest on the headland but had become a swift and expert flier, inferior to his parents only in experience and, more noticeably, in endurance. Both full-grown ospreys could remain on the wing from dawn to dusk and seemed to be as much masters of the air at dusk as they had been in the morning, but the young ones always sought the headland after making a few dives for fish and there they would stay until they had recovered their strength.

However, the knack of actually catching fish continued to evade them. Every other part of the process they were able to carry out. They had learnt how to fly slowly over the waves at a height of about thirty feet, they had learned how to intersperse their flapping flight with intervals of gliding, they were able to detect fish swimming in the green, translucent depths, they were able to dive down after them and then, having missed, they were able to shake the water from their plumage and rise shining wet, the spray misting from their feathers.

78

Nor was any of the timidity that had marked their initial experiments in flight present in their efforts to make a kill. On the contrary they were bold in the extreme, almost reckless, hurling themselves down whenever they sighted a quarry and always vanishing below the surface.

Probably many of the fish they tried to kill were out of reach, and perhaps the position of the quarry that was within reach was badly misjudged. Every fish they saw could hardly be at the maximum depth yet from the violence of their plunges into the water this might well have been the case.

During this period the parent birds continued to feed them and even after the learners had started to catch fish for themselves the feeding process was continued for a while, otherwise they might have starved; for their repeated plunging dives and strenuous emergences were tiring and a good part of the day had to be passed in resting.

However, as their skill gradually increased so did the supply of food from their parents gradually diminish, until the relationship between the four had become almost as casual as it was between the older birds and other members of the colony. It was in these circumstances that a series of events took place which, though perfectly in character in themselves, yet led to an extraordinary clash between the ospreys and those enemies who had robbed them of so many of their catches, the frigate birds.

Every member of the colony had left the neighbourhood of the headland for hunting grounds somewhere up or down the coast, the only exceptions being the two young ospreys, both of whom had killed earlier in the day and were now sunning themselves on the edge of the nest. They were, of course, much younger than any others of that season's hatching.

Carrying what would have been one of the last fishes she would pass on to her rapidly growing offspring, the female osprey suddenly appeared over the rim of the cliff. She dropped the fish in the centre of the nesting platform and, after a short period of preening her feathers, also settled down in the sun. Evidently, she too had fed recently.

The somnolent trio's rest was rudely disturbed by the attempt of a frigate bird to filch the unwanted fish. But the prize had slipped down between two of the dead branches that made up so much of the nest and the big black-plumaged bird was unable to pick it up cleanly. Taken unawares, possibly for the first time in her life, by the swift silent approach of an intruder the startled osprey struck out at random

79

and knocked the hovering frigate bird out of the air. For a few moments the great sea hawk lay sprawled across the nest then, scrabbling desperately with its tiny feet, pushed itself over the edge and was flying again, pursued by one of the young ospreys. The parent osprey also bounced into the air but hung there for a while, watching her offspring in his hopeless chase.

Noting the disturbance and observing, too, that the attention of the ospreys was not upon the fish lying so enticingly in the open a second frigate bird dived, but was unable to pluck the prize from between the sticks before the female osprey descended on him. Next moment the sky seemed to be darkened by hosts of frigate birds, though in fact there were only five newcomers in addition to the two already involved.

They were intent on killing the female osprey and they were so swift and pressed in so fiercely that the work was soon on the way to completion. Certainly the two young ospreys were quite ineffectual and did no more than wheel higher and higher over the combatants, shrilling their alarm notes with all the power at their disposal. It was probably the best course for them to have taken, for their agitated behaviour at the altitude they were attaining would be noticed by others of the colony. Meanwhile the swirling cluster of black wings could not be seen at more than a short distance, for it was now close to the water as the victim of the attack sank beneath the weight of numbers. The osprey had long ceased to call and was past the point of being able to offer any defence, splashing at last into the sea. This situation was not in itself serious as the osprey could have risen again as soon as her persecutors had gone. But they did not go, seeming determined to make an outright kill by repeated pecks on the head and neck. They floated above the object of their fury, each one darting in every few seconds to aim yet another blow at the almost unresisting osprey.

There was one factor, and only one, that prevented the kill being made before the return of the male osprey and that was the frigate birds' fear of the water. Ordinarily, they would descend to within inches of the wave-tops to pick out a dead fish floating on the surface but the osprey, in spite of her desperate position was yet an enemy who could strike back. Therefore, the possibility of being precipitated into the sea was greater than in more usual circumstances. So they were cautious as well as ruthless and remained acutely conscious of their enemy's long wings even though these were now merely paddling feebly in the water in an attempt to keep their owner's head from sinking.

The arrival on the scene of the male osprey was not the start of a dramatic rescue. But at least he swooped, screaming belligerently, over the swarming frigate birds and that was enough to create a diversion. Being above them and they so close to the water he was a serious danger and they were soon flying at his level, whereupon the situation changed once more. Well up in the air, the frigate birds closed in.

It was not their usual habit to launch such murderous attacks on ospreys but rather to harry and bully them into yielding up their catch, and it was the misfortune of the last arrival on the scene that the aggressive spirit of the raiders had already been raised to white heat and that he was one against many.

They flew over him and round him and under him. If he spired they were above him, if he dived they were faster than he, if he wheeled or dodged there was always a frigate bird to bar his way. His cries of anger gave way to shriller cries, with something of alarm in them. Instead of matching move with move he began to waver. They drove him harder; as in the case of his partner the speed, boldness and severity of the onslaught increased at the sign of his alarm. In appearance he was as confident and as angry as before, with crest upraised, beak parted with the effort of his cries now furious, now terrified, with his talons often menacing, or seeming to menace, his enemies. But gone were the fine, dashingly executed sweeps and towering ascents, missing the threatening lunges at one or other of his foes. He was merely beating his wings agitatedly and because he often remained defensively in the same place the talons that had loomed so formidable were now easily rendered innocuous by enemies who made their assaults from every direction except the front. For they were as adept at displays of bluff as he and some of their number kept the attention of the quickly-tiring osprey by threatening one head on collision after another.

Soon he, too, started to lose height and as he was beaten steadily downwards the sea hawks gathered more closely over him. Yet, every now and then, one would break away to hurtle skyward then swoop past the cluster of wings in a curve that must have been a quarter of a mile in length and which seemed to achieve nothing.

The hour was late, the ocean dark—almost purple. The spray under the cliffs was like blue smoke, the trees on the hump-backed western hills black palisades against an orange-red sky, and the ocean eastward already bounded by evening. Threshing her way free of the waves the female osprey set out desperately for the shore as her consort was forced down a short distance further out to sea. With her plumage sodden from long immersion she was flying in the lumbering manner of a swamp hen rather than an osprey. Slashing a dark streak of speed across the sky a frigate bird struck her down again. Nearby

82

a squalling company of avian swashbucklers was intent on another killing.

But from the north and from the south, some flapping in at low level, some sailing unhurriedly out of a sky now more purple than blue, the ospreys were returning to their roosting places on and around the headland. Most of them came like toilers after a day's work, these skilled catchers of fish. The young ospreys of that year came too. Some were strong fliers and all were older than the two young ones who had never ceased to soar high above the centre of activity shrilly crying their alarm.

Soon there were more ospreys in the area than frigate birds, and as both ospreys and frigate birds were particularly wary of fighting against odds and always preferred to make sure that the enemy was heavily outnumbered before joining battle, a distinct change came over the scene. Relief for the male osprey came quickly as two ospreys, reckless with fury, hurled themselves at their kinsman's assailants. Then with crests upraised and lethal talons full-spread, they veered away and rushed skyward as though gaining elevation for a downward assault. The bluff worked, the cluster of black, long-winged birds suddenly bursting apart and its fragments whirling into the sky. Within moments most of them had been swallowed up in the gathering dark. Most, but not all. For some reason three of the frigate birds had not used their peerless speed to escape from what would probably have developed into a crowded game of thrust and parry rather than a fight to the death. Perhaps they had gone back for a final assault on the exhausted osprey or they may have believed that the ospreys, in spite of their warlike display, would never come actually to grips. Whatever the explanation the three sea hawks were now surrounded.

They were not only much bigger than the ospreys but swifter, though not so swift that they could slip through the hollow sphere of their envelopment before some of the ospreys, by travelling a shorter distance, could intercept them. Every moment more ospreys were being drawn into the pandemonium until every member of the colony was whirling around the sky in one layer or another of a flickering, dancing orb of wings—the older and most aggressive birds closest to the three prisoners and so on to the outermost stratum of screaming, gyrating youngsters.

The clamour expanded to stunning proportions. Besides the osprey there were now terns, gulls and tropic birds rocketing through the night sky—for it was almost dark—and every bird was screaming, or whistling, or calling as loudly as it was able, the noise causing more excitement, and the excitement causing more noise until the level of frenzy was such as only a mass concentration of seabirds could reach.

The frigate birds were in desperate plight. Their favourite form of offence, a lightning swoop, a blow, was denied them now that their mobility had been contained and they had only their beaks for defence.

So they in turn were being pressed down, steadily, remorselessly. Then one of the three, more daring or more terrified than his companions, flew straight at the massed ospreys and got through. Some of the ospreys gave chase but the fugitive was too fast. However, the wall of the sphere had been breached and the two remaining frigate birds flashed towards the opening. Both were struck, for black feathers spun in their slipstreams, but both gained the outer air. One of the fliers vanished into the dark, the other flew erratically towards the inland hills. An osprey ranged alongside and, simultaneously, the black wings of the sea hawk sagged and, gashed by the talons of one of the ospreys, the great bird fell into the trees far below.

It was some time before the ospreys quietened. Then one by one, they dropped down to their resting places for the night. Three of the ospreys roosted on the nest on the headland—two young birds and a full-grown male with wings drooping from exhaustion.

Next morning was as many mornings; a calm sea, a dreaming coastline. The three ospreys on the headland took wing before the sun had cleared the horizon. Gannets were fishing near the submerged reef, myriads of smaller birds wheeled and soared above the islet, high up the frigate birds waited.

A big blue-grey and white eagle sailed across the headland on broad unmoving wings and followed the line of the beach, searching the sand and the shallows. He swooped low and hovered over a bundle of feathers stranded by the waves.

A gannet, full-fed, rose heavily from the water and made towards the islet, a black-plumaged bird with a forked tail and a tremendous spread of wing came hurtling down from nowhere to intercept. The gannet croaked despairingly and increased the rate of its wing beats...

84

# Red Kangaroo

THE quiet of the plains, the level distances broken only by a few
solitary trees, the clarity of the air itself created an impression
of infinite space. There was no cloud to check the flooding light,
no moving shadow to mar the bland passivity of the day; and the
black mote that was a high-flying crow seemed to hang motionless
between earth and sun. Its harsh, far-carrying cry pierced the silence
smoothly as a needle slides through silk. The plains were like a great,
calm sea, and whenever the wind blew a succession of waves, faint but
perceptible, would sweep through the dry and silvery grass and drift
away gently into nothingness.

Five kangaroos were grazing in the open not far from one of the
infrequent wilga trees. At intervals one or other of them would sit erect
on his haunches to search for sign of danger and, whenever the full-
grown buck towered out of the grass to peer around, test the breeze,
and twitch his attentive ears in the manner of a nervous horse, he
was seen to be much taller and heavier than the three blue flyers, as a
man is to a small boy in comparison with the youngest of the five. Indeed, so great were the differences between them, both in size and in
colour, that it was hard to believe they were all of the same family.

By reason of his superior height the big red kangaroo should certainly have been able to see further than the others but probably he
was not as keen-eyed as the does or as quick to detect anything unusual.
However, he was more cautious and if danger threatened, he never tarried
as they so often did, hopping lithely up and down in the same spot with
their heads screwed round and their eyes sparkling with interest as they
watched a horseman or, perhaps, even a greyhound, speeding towards
them. On such occasions he always made for the hills at once and with all
speed, thumping earnestly along as fast as his great weight would allow.

85

FIERCE ENCOUNTER

About mid-morning the red kangaroo idled towards a wilga tree
and sprawled out in its shade, propped on his elbow, his long hind
limbs crossed and a grass-stem in his mouth. Sometimes, he scratched
his ribs with his free paw so that, except for his tail, he might have been
a very tall man resting. The blue does and the young buck came to lie
down beside him.

Vigilance now seemed to desert them utterly, probably because
they instinctively knew themselves to be virtually invisible as they lay
there in the shadow of the tree.

The day was past its hottest when a horseman turned off the track
winding through the paddock from the main stock route. His course
would have taken him a fair distance to one side of the wilga trees and
if the kangaroos had kept still they would most likely have remained
undisturbed. But, at the mutter of hooves on the hard earth they
bounced anxiously to their feet, whereupon the man, bored by miles
of solitary ambling, sooled his dogs at them. Though half a hundred
kelpies could not have torn down the big kangaroo, even if they could
have caught him, it was he who led the flight, closely followed by the
young buck. Naturally the inquisitive does stayed. From time to time
they made preoccupied little jumps after the bucks but most of the
time they simply stood with their heads screwed round the better to
watch the approaching kelpies, now fast closing in.

At last they, too, decided to go, effortlessly keeping ahead of their
pursuers until the man startled them with a shout as he called back his
dogs. For the space of two breaths, the blue flyers streaked away at a
speed that explained, in some measure at least, their refusal to take the
kelpies seriously. Then they stopped again, to scan the retreating dogs
and the man and the horse until, their interest satisfied, they were content to stretch their slim and beautiful legs along the track taken by the
bucks.

Early morning found the kangaroos working back the way they
had come. The grass was richer in the paddocks closer in and soon the
little company was grazing with a mob of sheep not far from a homestead. When the sheep sought the shade trees so also did the kangaroos.
That evening they were not disturbed by horsemen or dogs and at
sundown they headed for an earth tank in the corner of the fence. As
they approached the drinking place the biggest kangaroo dropped
back, his head turning watchfully this way and that, his nostrils testing the breeze.

The plains were hushed. Grass stems rustled faintly in the light

86

breeze, a few sheep filed silently over the earth mound around the tank and passed from sight, and there was no other movement. The doe nearest the water, half-seen in the dusk, hopped quietly up the slope, paused in silhouette against the pale green band of light above the horizon, then moved unhurriedly down the other side.

Behind her came the other two blue flyers, followed by the 'old man' and, lastly, the youngster. Slowly, suspiciously, the bucks approached the water and, at last, began to drink.

Shy though they were in many ways the kangaroos showed considerable boldness in reaching the choicest patches of grass, even to the immediate environs of the homestead with its guard of dogs, and often they would retreat only when the uproar of barking and the rattling of chains had reached an intimidating volume.

As the season advanced the country round about dried up. Grass withered and was blown away until there was more bare earth than

pasture. Other mobs of kangaroos moved in from drought-stricken areas to the west and south, and on most evenings there were almost as many marsupials as sheep around the earth tanks. In spite of the sparseness of feed thereabouts the young red buck grew bigger and heavier with every week, and his constant hunger made him as daring as any.

Even when the stockmen carried rifles the kangaroos remained, merely moving round the menacing figure of a horseman whenever they saw one. One of the blue flyers was shot because her curiosity caused her to wait when she should have fled and more than once the young red buck heard the whine of a bullet.

It was about noon on an especially hot day that the two remaining blue flyers in the little group caught sight of a horseman cantering towards them. Immediately, they bounded to their feet and, as was usual with them, stayed to watch while the two bucks set off with long, urgent leaps. But, directly in front of them, there appeared two more horsemen. The red kangaroos swerved away at an oblique angle. Other kangaroos were on the move, all now travelling on a course running somewhat towards the range of distant hills. Ahead, and to the right, other horsemen were riding. As the company continued on its agitated course its numbers swelled until there were kangaroos leaping in hundreds along the paddock's boundary.

Dust swept up and hung above them and the day throbbed with the thumping of many pads on the hard ground. All colour was lost. There were no red kangaroos, no blue flyers but only a grey mass of terrified creatures milling along a fence. Hesitant, balking, hanging back, the press was forced relentlessly on by the horsemen who now formed an unbroken line behind them. A second fence, newly erected, had appeared on the other side of the trapped animals. The fences were angled towards one another. In front was a belt of dense trees with an opening between them. From the trees, sometimes, came the glint of metal catching the sun. But the fences, reassuringly, had ceased to close in.

Still at the side of his bigger kinsman, the young buck bounded unwillingly along the way that all, it seemed, must take. He was jostled by others in the fog of dust as panic now gripped the driven kangaroos. Some of them crashed into the fences, others tried to leap over but, generally, forgot to reckon with the top wire and, their hind legs passing underneath and their bodies swinging over, they hung there to be dispatched later and at leisure by the hunters. Some of them, certainly, cleared the obstacle for the height of it was well within their leaping powers and it was only the blindness of terror that brought most jumpers to disaster.

The vast majority, however, feared the fences as much as they feared the hunters, who so far had shown only determination to drive their quarry on. The young buck never had any clear idea of what happened when the great herd of kangaroos reached the wilga trees. There, it seemed the world was coming to an end. Concussion after concussion blended into a single deafening roar. Scarlet flashes stabbed from under the dark trees as the guns mowed down the bewildered, milling mob. The dust haze thickened until it had turned the sun to a dull red glow.

The old red kangaroo went down with a shattered hip and then a second charge tore mercifully through his chest. A group of blue flyers, galvanised at last into action, flashed by with the speed of swooping birds. All except one tumbled in succession to the ground and lay still, the survivor vanishing in the gloom beyond the trees to safety. The young red buck followed and, being in the midst of a gush of terrified animals, also made his escape though almost every other kangaroo was shot down.

Only a few lived through the ten minutes of nameless terror; of the young red buck's former company there was no survivor other than himself. The two blue flyers were bloodied heaps amongst the hundreds of red and blue mounds of fur dotted from one end of the passage through the trees to the other.

So the young buck left the fatal plains and their sere grasslands to the grazier and his stock and made for the drought-stricken hills. For three months he lived by dismal dawn to dark foraging among sun-heated rocks on steep slopes where there was no growing thing except for a few eucalypts—and their leaves were almost always out of reach. During that time the rocks were never cool, not even at dawn. All night long the warm, dry breezes stole about under the stars and the scent of dust was always in the air.

The red kangaroo was himself as restless as the wind, for the unease of a starving animal is full of weariness, and the gaunt glassy-eyed creature nosing at the parched ground and moving painfully, yet with a kind of bemused determination, upon its ceaseless search for food was quite unlike the red kangaroo which had left the plains. A wisp of dried grass close to the side of a rock, some stems blown against a tree, a leaf or two low enough to be reached by an upthrust muzzle, a few burrs or other seeds picked out of the dust—these were sufficient to keep him dying more slowly than would have been the case if there had been nothing to be found.

When the rain came at last the red buck was young no longer. Also, he was no more than a skeleton held together by what looked like a skimpy, rust-coloured mat rather than a living skin. His frame was full of hollows and protuberances; his hind limbs had wasted to bone, skin and sinew, and there was no alertness in the cast of his gaunt head.

As it happened the drought broke first on the plains, and there the red kangaroo returned without delay, often grazing close to the homestead and visiting the big earth tank to drink at sunset. Once more the great hole scooped out of the ground was full of sweet water and the company of kangaroos now visiting it was very like that of which the red buck had once been the youngest. Now he was a leader, almost as tall and heavy as the red kangaroo the shooters had killed on the terrible afternoon of the drive. He was not as old, of course, but his long-boned, rangy frame was already that of a very big animal. Three blue flyers followed him wherever he went and there was a young buck, too, with no ambitions at present and no desire but for company.

While the good seasons continued the kangaroos never moved far from their favourite pastures. On hot days they would feed in the early mornings, then drowse in the shade until well into the afternoon when they would begin to graze again; in winter, when the rain was being swept across the plains by a cold wind, they would lie in comfort, almost in luxury, under the lee of an ensilage stack close to the homestead.

They were harried at times by the station dogs, but were only once really in peril and that was when a rabbiter loosed a pair of greyhounds against them. Immediately, and exactly in the manner of his predecessor, the big red buck set his course for the hills. And the reaction of the does was just as predictable. They retreated a little way, then stood watching, each with her attention focused upon the man. Nor did they move when he brought two lean, brindled dogs from the rear seat of his car and held them leashed. The dogs peered this way and that, excitedly. They capered and danced tip-toe, pulling at the leathers; but the man held them and tried to calm them, though they bowed their sinewy bodies and wriggled and stared about, knowing there was game to be hunted but seeing nothing. The man pointed, the dogs gave no sign of sighting the quarry for the does were utterly still, and it is movement and movement alone that attracts the eye of the animal.

Then one of the flyers craned higher and the dogs lunged, spurting dust in the man's face. The leashes fell, the brindled back-bones whipped double and stretched, the narrow long-jawed heads ducked and rose and the greyhounds were racing. Straight as flung spears they flashed over the intervening space. The blue flyers bounced agitatedly up and down on the same spot and stared incredulously, stupidly. Three smoke-blue forms then leapt away, three smoke-blue shadows skimmed the silvery grassland. The man shouted in triumph and admiration.

So they flew, the three beautiful does and the sinewy hunters. The dogs must catch them quickly or not at all. They would have killed either of the bucks from that distance but the does were swift, their leaps so long and low that they, too, seemed almost to be running. The dogs made little impression on the two younger flyers but the oldest was nearly overtaken in that initial surge of speed. They turned her— the oldest fugitive—and the second dog missed with a leap that took him far beyond his mark. The blue flyer swung in a new direction and came round, almost back over the ground she had already covered. Running forward the man shouted and waved his arms but she kept determinedly to her new course, heading for a wide patch of reedy grass. There was no water in the place but the grass was tough and long. The fugitive sailed across with undiminished speed, the dogs swished into it, then tried to leap over it, bound after strenuous bound, while the blue flyer fled away as though they had been standing still. She swung again in the direction the others had taken as the dogs burst out of the cloying grass and began to race again. But their magnificent fire had burned out and the blue doe continued to draw ahead. The quarry was lost. Let the dogs trail on, their chance had gone.

The man whistled. Both dogs lost speed. They cantered a short way further and stood watching the small, blue-grey kangaroo bounding, leisurely now, away into the distance. They sniffed here and there about the ground and came trotting back to their owner, their lean flanks pumping and their tongues hanging from their parted jaws.

Next day the kangaroos were back near the homestead. Indeed, the big red buck had become a privileged personality about the station, where every stockman knew him because of his great size. They said, truly, that he was a comparatively young animal and that he would be a giant in a few more years. They were proud of him, even if he did eat as much grass as a bullock. They never sooled the dogs on to him, they never shot at him or at any of his group and they warned the younger men about the place to leave him alone. They reprimanded the rabbiter

91

when he remarked that his dogs had missed a kill only by a piece of bad luck and told him that the big kangaroo and his following must be left undisturbed.

So the red kangaroo became bigger and heavier with each passing season. Five or six years after the drought which had caused the death of so much stock another long dry spell set in, though a number of thunderstorms saved the red kangaroo's range from utter bareness. In fact, it became an oasis in the midst of desolation; and such was the invasion of native animals that, again, plans were made to kill them off in order to save the sheep.

Careful instructions were given to every shooter, however, that the big red kangaroo must not be destroyed. It was maintained that there would be no excuse for failing to recognise him. Some bushmen said he stood eight feet high in his socks and must weigh close on twenty stone but they were carried away most likely by the fervour of their admiration.

Anyway, it was indeed a fact that there was no mistaking the great creature when he came thumping along in the midst of terrified hundreds of his kind. As before, the guns roared continuously, the horsemen shouted and urged the victims into the lane of death; as before, the brown earth was littered with red and blue-grey corpses while dust and smoke and the bitter stench of gunpowder rolled slowly away over the plains. Through it all lumbered the big red kangaroo. Dazed, stupid with the uproar, he bounded on, sometimes stopping, sometimes leaping forward again in a sudden access of terror that drove him for a space at a good speed.

And he remained unscathed. No charge of gunshot passed within yards of his red hide and, yet, he turned all of a sudden and leapt towards the wilga trees. The uproar ceased. Somebody roared jeeringly 'Look out, Joe. He's after you!', just as the heavy animal, blind with fear, charged straight at the youngest shooter in the party. In self-defence, as he said later, he fired point-blank and missed. Next moment he was knocked spinning to the ground and, as he fell, the second barrel of his wildly-swinging gun brought down a shower of leaves and shocked ten years off the life of the man standing next in the line. As the red kangaroo crashed away through the deadwood under the trees the guns returned to the task of claiming the last forlorn cluster of victims.

But the incident served further to expand the reputation of the big red kangaroo and the same men who had maintained the boomer was eight feet high also declared that he had not been fleeing in blind panic at all, but had made a furious and courageous attack upon the destroyers of so many of his kind.

There were few survivors of that second holocaust through which the big kangaroo had passed unhurt, and by that same evening the tiny remnant of the host was far away in the hill country, where nothing grew except the tough bush trees.

Yet before many weeks had passed a few kangaroos had returned to their former territory. The red buck was one of these, bigger than ever—more picturesque than ever, too, from a distance, but ragged and scarred when seen close at hand. He was bolder than before and sometimes disdained to flee from the station dogs. Rather would he wait, bounding a short way in retreat to begin with and then, having drawn them away from the station buildings, he would turn and make rushes towards them as they clamoured in a semi-circle about his towering form. His truculence became a legend in the district and many a bushman came from round about to see the red boomer of Merriba. For boomer he was by this time, the biggest in the land. Not even the old-timers—not even the finest liars amongst the oldtimers—could tell of a red kangaroo to approach the size of him. And he continued to grow, as such animals do, even in old age.

Accompanied always by two or three graceful little blue flyers, and, perhaps, a few of his half-grown offspring, he would drink every night at the earth tank, which was his favourite spot, graze in the best paddocks, lounge under the wilgas and go hopping ponderously from one patch of good grass to another exactly as he liked. The time came when he would not bother to get out of anyone's way, and a story was told of his sprawling across a track and refusing to bestir himself even when the mailman drove along in his motor lorry. The tyre marks were there to see if you didn't believe him, the mailman said. You could see where the lorry had been obliged to swerve and, at the same point, you could see where the red boomer had been lying in the dust.

It was when the station was shearing that the red kangaroo suffered a jolt to his dour self-importance. For a shearer who was also a grey-hound fancier had brought two dogs with him. They were valuable dogs, although the squatter maintained that such beasts weren't worth feeding as far as usefulness was concerned. In this he may have been correct, as the shearer may also have been when he expressed the opinion that when it came to chasing kangaroos the station kelpies and cattle dogs, for all their wonderful ways with stock, were themselves not worth a bumper.

Having been warned about the special status of the red kangaroo the shearer set out one Saturday afternoon to see if he could set his greyhounds at some other game and it was unfortunate that he blundered on the very animal he did not wish to meet—the big red kangaroo with his retinue of blue flyers and younger animals. In a moment the greyhounds were in pursuit.

The blue flyers swept away, followed by the younger kangaroos, but before the red boomer had taken more than a few bounds the dogs were up with him. But they did not close. There was something too menacing, too confident about the great red form with its back to a tree, its forepaws ready to grab.

While the owner tried to call off his greyhounds a stockman was yelling indignantly at him for daring even to disturb the old kangaroo. But, as he shouted, his own dog raced in and, for the reason either that it was more courageous than the greyhounds or less knowledgeable, attacked headlong. Reaching down the great kangaroo seized the kelpie in his forepaws but, before he could disembowel the squirming animal, the stockman jumped from his saddle and picked up a heavy stick. Brandishing it, he advanced. Tossing the kelpie aside, the kangaroo waited. The man continued to go forward.

But the red boomer seemed taller when seen from ground level. He swayed on his great, heavily-muscled hind limbs, too, as though contemplating a sudden bound and his near-human forearms, corded with sinews, made menacing, clutching movements. The man hesitated, the kangaroo bounced resiliently up and down, his front paws spread; the dry leaves crackled under his gently-thumping pads.

'I'd watch him if I was you,' the shearer warned in a flat, but tense, monotone. 'If he grabs you it'll be "goodnight".'

At that moment the kangaroo bounced forward, the stockman jumped back, tripped, and fell. He glimpsed a red shape looming over him, then the greyhounds dashed in and the kangaroo was bailed up again, cornered but defiant.

The stockman clambered to his feet. He was not a timid man but he was a discreet one. He was a man, too, who recognised real danger when he saw it. Also, like most men, he hated to be laughed at and he was not pleased with the shearer's grim amusement.

'A bullet would be better,' he said. 'See if you can keep him there. I'll whip back to the homestead for the thirty-two.'

But he had no sooner remounted than the red kangaroo, sensing a lapse in the concentration of his enemies, set off with long, thudding bounds. At once, the greyhounds were again in pursuit, overhauling the fugitive without difficulty until the chase veered in a new direction, the dogs suddenly finding themselves blundering through a litter of deadwood in the shadows of a thicket of trees. Bounding on triumphantly into the clear the kangaroo thumped on his way towards the earth tank and was almost there when the greyhounds swooped across the open ground like two huge swallows and turned him. One of the dogs threw the kangaroo off-balance by hanging on momentarily to the butt of his tail and the great creature slithered sideways in a cloud of golden dust just as the second dog sprang. The attack was deadly dangerous but missed the throat, tearing a deep furrow in the kangaroo's shoulder.

Then the stockman tried to end the battle with his stirrup iron. If he had had an older horse instead of a brainless, nervy youngster under him the matter would have been settled there and then, for the full swing of the iron would have stretched the great creature in the dust or, at least, reduced him to such a state that the dogs would have been able to complete the kill without further trouble. But an unpredictable mount is no aid to confidence and the rider had no wish to be pitched into the ready arms of the red boomer. Resolutely, with one dog now fastened to his flank and the other harassing him from the rear, the kangaroo forged towards the tank, over the earth mound surrounding it, and into the water.

It was stalemate. The greyhounds splashed round in the muddy shallows, panting, and the horseman, too, stood at gaze. The owner of the greyhounds arrived, perspiring and out of breath.

'Dive in and finish him off with your stirrup iron,' he suggested drily, calling back a dog which was venturing too deep.

'Yair,' the stockman agreed, more unappreciative than ever of his companion's idea of humour.

Muddy wavelets lapped the kangaroo's chest as he returned the stares of his enemies. A fleck of foam dropped from his mouth and his eyes were wide from the efforts of the last few minutes; the slash across his shoulders had had time to bleed and, also, at this close range, the boomer's skin showed many more scars of past battles and mishaps.

'Biggest 'roo I've ever seen,' the owner of the dogs observed, squatting on his haunches and pulling the makings out of his hip pocket. 'What do you aim to do?'

'Well'—the stockman drew a deep breath—'Well, I reckon I'll go up to the house for that gun. He could be a killer, judging from the way he came at me. You wait here an' keep an eye on him.'

The other lit his cigarette before answering.

'Right,' he agreed.

So the stockman wheeled his horse and urged it forward. Then he reined and, turning in the saddle, looked at the red kangaroo again, noting the signs of exhaustion and the ragged appearance of the animal's coat, or as much of it as was visible above the water. The dogs, still excited, but quieter now, rocked gently back and forth and their red tongues lolled, dripping saliva; the older dog yawned and flopped down in the mud.

'It's a bit far, the homestead. I don't think I'll worry about the gun,' the stockman said at last. He should have killed the red kangaroo while his blood was up.

Without answering the other man tossed a stick at the dogs.

'Come on,' he said. He put them on their leashes.

They returned the way they had come. Before they had gone very far they heard a loud splashing in the dam. The greyhounds tried to break away but their owner held them; prancing and sidling along on his skittish young horse, the stockman nodded approvingly.

They went on in silence.

# *Wandana*

THE black swan floated motionless, watching and listening, her scarlet bill gleaming in the fading light, her long neck vigilantly craned. But there seemed to be no sight more disturbing than a few ruffles of deepest blue hurrying before the breeze or, beyond the dunes separating the lagoon from the sea, a flight of gulls drifting down to the beach. The only sounds were tall reeds rustling and the drowsy roar of surf.

Yet despite the calm of evening the black swan fed for only a short time then, hurriedly, as though she had already tarried too long, made back the way she had come, her great, white-tipped wings flaring in the dusk as she climbed the steep side of her nest near the lagoon's shadowy edge.

She had been the sole protector of her brood since the disappearance of her mate a few days before, and had not dared to venture far from her eggs. Next morning, however, she was certainly not alone, for when she launched herself into the water soon after dawn she was successively trailed, surrounded and impeded by four active, shrill-voiced young ones.

The sun was lighting the crests of the sandhills when the small company struck boldly across the lagoon. Bobbing perkily upon the wavelets and showing that strange, instinctive awareness of their surroundings which is the instant heritage of so many bush creatures, the cygnets forged energetically.alongside their guardian. They paused in the shallows where the breeze was hissing amongst the reeds, flexing them like bright rapiers, and there the swan wrenched and dug with her strong bill while the cygnets darted at the fragments of broken roots and greenstuff that floated to the surface.

As the heat of the sun increased the little group paddled back towards the nest where the cygnets sought the cool shelter of the reeds. And, whenever one of them set off deeper into the mysterious thickets that beckoned with glints of water and moving patterns of light the call of the older bird brought the adventurer back to the outer fringes and nearer to herself.

The brilliant day burned on. A cormorant fishing in the centre of the lagoon dived at intervals, always to reappear in approximately the same place, its white throat and black head distinct against a background of blue water; behind the cormorant, as an unbroken bank, were the green reeds of the opposite shore, and above the reeds the white glare of sand. The world seemed asleep. Even the skirring of insects had died away and the sound of the sea was a far-off murmuring.

The black swan began to feed and it was when her head was thus submerged that a man appeared over the crest of one of the higher dunes. Regaining her normal swimming position the bird looked round and about on every side as was her habit but the newcomer was now as motionless as a blackened tree trunk, or a slab of rock. Tall and bearded, armed with spear and woomera, he fitted into the scene as perfectly as the drifting clouds. Again the swan thrust her head below the surface and when she next came up to breathe the man had gone.

At about sunset that day a cloud rose out of the sea, and by midnight rain was murmuring in the water round the nest and pattering on the resilient sheath of plumage which protected the swan and her brood from rain and wind and wave alike.

Dawn was scarcely worth the mentioning though it did enable the Aboriginal warrior who had marked the nest on the previous afternoon to see his way ahead.

With infinite care and patience he had travelled the length of a narrow strip of heathland between the ridge and the lagoon until he came to a point where he judged he would be opposite the nest. There in the wan dawn's glimmer he paused to wipe the rain from his throat and to rest before stealing into the reeds. He could see the isolated cluster of sedges which overtopped the surrounding thickets of reeds. The nest was beyond that mark, but not far beyond it.

Crouching, he made his way forward. The rain dashed into his face and the wind whirled away his scent. Faint glints rippled along his arms and shoulders, and his spear glided inaudibly behind him. It was possible at last to steer by the cluster of tall sedges without raising his head above the level of the reeds and so, step by painful step, making even the movement of a finger the subject of thought and of conscious muscular control, he stole forward until he reached the place from where he would strike. Though he had not sighted his quarry since the day before he began to make ready with absolute confidence. Drawing the spear somewhat past him he fitted the shaft to the woomera. Then he planted his feet so that, on rising, he would be in position to throw, and slowly straightened his body.

So quiet, so deliberate, so well-concealed was the Aborigine that it would have been remarkable if the swan had noticed him. Standing erect, being able to see the swan through the tips of the outermost reeds and, having his spear aimed, poised and vibrant, the kill was as good as made. The quarry was too close now to have any chance of escape. Such a large and heavy bird could never outstrip a flung spear, and the range was so short that a miss would be disgraceful.

The black swan must have seen or heard something suspicious, however, for her head flicked round and she had partly raised her wings, either in defence or for flight, when the shaft transfixed her.

A foamy path burst from the reeds as the man crashed forward and seized his prey. He made as though to withdraw the spear, then changed

his mind and, using the shaft as a lever, swung the dead swan to his shoulder.

Discovered, the cygnets raced from the nest. The man made a grab but missed. In any case the chasing of such small game was hardly worth his while. Later on, perhaps, he would send someone to search the reeds, though here again the fledglings were so small that their capture would scarcely repay the trouble of telling the women where to look for them. With the red bill of the impaled bird tapping at the small of his muscled back he waded across to the dunes where the other members of his tribe slept on the rain-drenched sand.

Within the reed-bed where it was almost dark the four cygnets hid in trance-like terror. Wandana, the youngest, had no clear recollection of the disaster, but the man-scent which had emanated from the killing itself would always signal instant danger. His was a reassuring hiding place, nevertheless, for the outermost reeds recoiled so resiliently before the wind that the force of it was deflected and in this manner the cygnets were surrounded by a tranquillity in complete contrast to the turmoil of the outer world.

It was the grey teal who next day enticed the cygnets from cover. The teal called in their soft, clear notes as they came scudding nearer in a long, downward slant, and the small disturbance made by their arrival had scarcely faded before the cygnets were edging towards them.

No cloud marred the brilliance of that morning. Grass on the sand dunes separating the blue of the lagoon from the blue of the sky seemed to have gained in freshness and there was richer colour everywhere. The lagoon sparkled as a cut opal, green in the shallows, ultramarine in the deeper parts, almost black over patches of submerged weeds and purple in the shadow of a ti-tree thicket on the shore. Sunshine glinted on the metallic green wing patches of the sober-plumaged teal as they swam closer, calling to each other.

Wandana was the most alert of them all, and the most apprehensive. He would take fright at even a harmless gull, provided it appeared with some degree of suddenness, and he was always leading the other cygnets in frantic dashes for cover. Generally, the teal would wait to see what had caused the latest alarm but sometimes they, too, would be caught up in the panic. The black cormorants, particularly, never failed to cause a commotion and often fooled the whole party. They came into view so unexpectedly with their long wings spread in such a hawk-like manner that neither teal nor cygnet would pause for a second glance, the cygnets thrusting deeper into shelter, the teal diving.

There were many birds on the lagoon that day. Another group of teal was feeding near the eastern bank, a heron drowsed one-legged in the shallows and a big cormorant who had just caused consternation by his dramatic appearance, scored a white streak across the blue water as he flapped heavily into the air again to join his kinsmen on the bank. There he stood motionless, his angular, coal-black wings spread out to the sun. The assemblage made a quaint picture for it included some little cormorants, about the size of wild ducks, as well as a number of the heavier black cormorants almost as large as swans. They were in every attitude—some twisting their snake-like necks this way and that to preen their feathers, some hunched up and immobile, some standing with their wings outstretched in a position that looked uncomfortable enough but which they held for long periods without stirring.

The cygnets drowsed in the afternoon warmth but for all their seeming somnolence the older birds remained vigilant; and when a wood duck on the far side of the lagoon uttered one sharp warning note and dived so, also, did the teal.

Wandana could see no danger but he did not hesitate, and four trails of spray licked into cover an instant before the falcon crashed into the reeds, his long sinewy legs and talons at full stretch. He had been scarcely a wing-stroke too late. As he beat angrily clear he threw the cormorants into a flurry of brandished wings by veering menacingly towards them, though he would have been ravenous, indeed, to stoop at such rank prey. In a few moments he was out of sight.

Notwithstanding such frights as this the cygnets spent more and more time foraging in the open and it was inevitable that this trend would continue for they were growing fast and their appetites were increasing in proportion. Before they left for the inland rivers the teal had taught the young swans something of discretion and had saved at least one of them from the falcon. The reason for their abrupt departure was the return of the Aborigines to the sand dunes.

Nomadic and restless though they were the Aborigines rarely moved away from any locality where food was plentiful and at that time the beach was full of pipis. These the women and children fished out of the sand to their heart's content, to cook and eat almost without pause, until the presence of the four young swans having been marked, and it having been decided that they had attained a size and weight which made them worth the trouble of taking, the warrior who had speared the parent bird now made a plan to hunt down her offspring.

The day he chose was boisterous and cloudy with a southwest

wind. Leaden-tinted waves sped diagonally across the lagoon and slapped upon the low, muddy bank where the cormorants loved to preen themselves in sunny weather. But on that day there were no watchers to give the alarm when in the gloom of late afternoon the hunter slipped into the reeds. He was carrying a heavy stone in one hand and a branch broken from a paperbark in the other. There were leafy twigs from the same tree thrust into his hair. Because the water was shallow and the reeds thick his progress was difficult.

Somewhere out of sight he knew other members of his tribe were watching—some admiringly, all of them critically. Prostrate, trailing the branch behind him, he stole towards the open water. He was in no hurry. As the water deepened he changed his attitude until he was in a squatting position, with only his head above the surface. The stone in his left hand kept him comfortably submerged and, on reaching the outer margin of the reeds, he rested on his haunches awhile to study the scene before him and to drift forward very slowly the branch he was holding in his other hand. His shrewd black eyes peered through the mass of fine leaves and the branch continued to glide against the wind.

Wandana and his companions were sheltering under the lee of some sedges, and none of the three noticed the gently-moving branch until it was mid-way between the two shores. Then Wandana craned his neck into the straight line of suspicion and, affected by his uneasiness, the other swans focussed their attention upon this object which had made headway so mysteriously. But the branch did not come any closer. It remained so long in the same place, in fact, that it became as any of the other branches snagged here and there in the lagoon.

Wandana fell to preening his plumage, his brothers resumed their feeding, the green branch edged nearer.

So observant were the eyes in the screen of leaves and so creature-wise the intelligence behind them that the swans, though occasionally curious, were never alarmed and made no attempt to leave their sheltered position for the wide, gusty expanse of open water. At last the advance ceased altogether. The branch was now very close to the four. Almost completely submerged in the shallow water, his left hand still clasping the heavy stone, his head concealed by the dense, fine leaves, the man waited. The wind swept the scent of him away from his quarry. There was neither sound nor movement except for the noise of the wind, the splashing of wavelets and rustle of the reed-beds.

The swans stopped feeding, and gently rocking in the calmer, sheltered water, tried to keep their positions by a few lethargic strokes of their webbed feet. Slowly, then more quickly, the wind eddies would push them into the open. The further they drifted the quicker the wind drove them. However, it was not until they reached the rougher water that these indolent swimmers, now about the size of wild geese, exerted themselves sufficiently to paddle back to the shelter of the reeds. Each time they were blown towards the centre of the lagoon they delayed a little longer before returning, and so drifted a little closer to the branch.

So strong was the man's grip that the branch might have been part of a tree embedded in the mud. And the fact that it had not been there, say, an hour earlier was of no significance. Nor, apparently, was the fact that it had approached against the wind. The swans had not seen it move, therefore it had not moved. It had always been.

The attack was a complete surprise. Incapable of rising quickly from the water as smaller birds might have done the swans were also panicked by the sudden apparition of a man almost within armslength; and the hunter quickly seized and killed two of the group. He was nearly successful in capturing a third. However Wandana and his surviving brother, just out of reach of his final lunge, managed to rise into the air and, driven by terror and the wind, shot erratically seaward with something of the speed but none of the grace of that enemy of their cygnet days, the falcon. They crossed the surf where the white foam leaped and the noise of it set their wings beating the faster. Ahead was an expanse of blue water wider than the widest lagoon, and no sooner had the swans put behind them the clamour of breaking waves than they wheeled landward again, striving for more height as they again approached the line of the surf. An inlet winding behind low hills was bright in a late sunlight as they planed smoothly down to alight between high and unfamiliar banks.

It was about dawn when from high overhead came a series of quaint, tentative notes, very like the creaking sounds made by tree limbs rubbing together in the wind. The two young swans answered and the voyagers swung in a wide circle, then flew on. Below and to the rear Wandana and his brother threshed the air in an effort to catch up. Although they failed to narrow the gap, and may even have fallen further behind, they were close enough to see the leaders go into the characteristic long glide that indicated the end of a journey. Wandana and his brother alighted some distance from the strangers.

The quietness of the inland was a new experience for Wandana. It might be argued, of course, that almost any experience would be new to him, he being so young and untutored. Again it might be added that, having had no parental protection after the first few days of his life, he had already been exposed to dangers which otherwise might never have threatened him. In one way he was less knowledgable than his associates of the same age—for some of the company were of the same season's hatching as himself—in another way he was wiser and warier; and every night, when the shadows of the river red gums were swallowed up in the dark, and the high banks with their fringes of grass were full of faint noises, the brothers were always the first to edge into midstream.

The young swans grew fast in the changeable weather before the onset of winter. Rain often dimpled the smooth river, and occasionally the wind whipped its surface into waves almost as boisterous as those which had roughened the more exposed coastal lagoon; sometimes the stars shone clear in a serene night sky, sometimes clouds scudded before a storm and the branches of the red gums strained under the blast. There were days too when the mild sunlight of autumn gleamed on the quiet water and the glossy black shadows under the banks might have been part of a painted scene—days when the only ripples were those made by the swans as they glided through patterns of light and shade. Wandana would have stayed there indefinitely. He was now the largest of the younger swans. His bill, formerly of a dark olive green, was turning scarlet and he was heavy enough and strong enough to lord it over his contemporaries.

If the decision to stay or go had rested with him he would have stayed. But then he was not yet subject to the instincts which motivated the older birds, and furthermore the decision did not lie with him.

The younger swans were thrown more and more into each others' company as the mating season approached and, whenever they did try to rejoin their elders, they were quickly driven away. The two mature cock birds were especially belligerent, each insisting upon having plenty of open water between himself and every other member of the small community with the exception of the hens. Then, on a night when the shadows cast by the moon were almost as emphatic as those of the sun at midday, the quiet was suddenly broken by the beating of wings as the whole company swept along the river, leaving a sparkling wake that ended abruptly as the fliers drew clear.

Wandana was last. Just ahead of him flew his brother, his great flashing wings thrusting him on into the brilliant night. The soft whistle of pinions and the mellow, creaking calls of the leader were now the only sounds as the phalanx rose between the double line of trees and mounted into the luminous sky.

The calling ceased. Long necks outstretched, wings rising and falling with rhythmic glimmerings of their white tips, the wild swans came round in a wide curve and continued to climb. Wandana moved more exactly into his place.

After the moon had set they flew on just as purposefully, and with daylight Wandana noticed travellers other than themselves. A group of teal, their short, sturdy wings blurred by the rapidity of their strokes swirled by. Flight by day was especially dangerous for them for that scourge of their kind, the falcon, would soon be ranging and waiting high above the air lines.

When the sunshine was shafting across the plains below the swans began a long descent towards a sweep of brilliant blue dotted with innumerable islets. The great swamp was covered with birds—teal, ducks, geese, grebes, widgeons, ibises, sheldrakes, and cormorants. Water hens, black as charcoal when in shadow but of a rich, iridescent purple in the sun, bobbed in flotillas amongst the overhanging canes of the lignum islets, a royal spoonbill waved his snowy wings as he rose from his nest and overhead the eagles floated in great, slow circles. Beneath the wedge-tails were the whistling eagles and the buzzards; lower down, a swamp harrier quartered the hunting ground for unguarded eggs or nestlings. Every now and then a multitude of birds would take wing together and wheel through the bright morning.

As the swans slid lower the din of the huge nesting and feeding ground came up to meet them, the full volume of that avian clamour expanding into a storm of noise so tumultuous that the call-notes of the swans were inaudible.

But if their thin cries had no effect it is certain that their right to a stretch of open water was uncontested as they came to rest in a glare of spray and a final flourish of wings.

Wandana paddled smoothly after the others. Yet, before the sun had reached its zenith, the older swans had driven away the younger. Vigilant, unsettled by the strangeness of it all, Wandana and his brother sailed on through the labyrinth, their black necks craned, keen eyes probing the depths of the lignum islets. Every bush contained a nest, and some had two or three.

Drifting on, down one twisting lane and along another, they found a mudbank where a profusion of plants grew in the sun-warmed ooze.

107

When he had fed Wandana splashed through the shallows, his heavy body swinging awkwardly with each step, and climbed the slope of the mudbank to its dry and rounded crest, where a pair of wild geese were sunning their beautiful, steel-blue plumage. Wandana fanned his great wings until the dust flew and the geese ruffled their feathers in anger.

Three grey teal drifted into the shallows and they, too, were in the undecked habit of immaturity. Unhurriedly, they placed themselves between the swans and the wild geese after the gregarious fashion of the aquatic birds. Last to join the group was another swan. He was older than the others though he had not mated that year. So he, too, became a member of the assorted band. But he was not as peaceable as the rest and often clashed with Wandana.

Gradually, Wandana was assuming an air of authority. He was seldom aggressive, however, and both he and the older swan generally remained on opposite sides of the group. Whenever they did come close together there would be, as like as not, the slash of a wing and a darting of long necks.

While the cycle of the rookery passed through its several stages the self-contained little company cruised the now clay-coloured waters, fed in the shallows, rested on the edge of the plain and took short flights from one feeding ground to another.

Eggs hatched in the uncountable nests, young were reared and escorted in clusters and in line ahead through the twisty canals in the lignum; in many cases, more eggs were laid as soon as the first family had been fairly launched.

Marauders waxed fat in that easy hunting ground. If a nest were left unguarded the owners might be sure that eggs or offspring would be gone on their return. Water rats glided through the abundant cover and accounted for many a fledgling, eels and goannas preyed upon the young birds as soon as they took to the water, and whenever a high-sailing buzzard closed his wings and fell out of the sky it meant that the contents of another nest would soon be nothing but a litter of empty shells.

Wastage of life from starvation, too, was great. Weaklings died young and the waters were always dotted with dead nestlings drifting slowly before the breeze until they became entangled in the drooping canes of the bushes. The wedge-tailed eagles fared well, as did the whistling eagles; and the most savage killers of all, the falçons, roamed the swamp as tirelessly as their counterparts had roamed the dunes,

108

ridges and headlands near the lagoon where Wandana had made acquaintance with the world.

There was little rain that season and as the water level slowly dropped, the teeming swamp began to show evidence of the size of its enormous bird population. Food was no longer plentiful; hundreds of nestlings, hatched late and deserted, lay dead in their nests and the water's edge was wrinkled with green scum.

So Wandana and his brother took wing, to fly westward as far as a wide river with trees along its banks and with underwater pastures on the inner sides of its curves, where the current was gentlest.

Autumn brought its mild and sunny weather, and winter brought its rain. Sometimes, when the showers were light, the swans would walk about the banks and graze as contentedly as wallabies, but when the falling drops were heavy enough to make the dead leaves dance in the grass they would often seek the shelter of the trees.

They stayed together, amicably enough, on the remote inland river, until the coming of Burrilda. It was about noon when she swished into the water upstream from the brothers. Although she had planed directly over them and though the black swan is wary and keen-sighted, both Wandana and his brother might have been invisible even when paddling near her, so little notice did she take of their presence.

Both cock birds courted the hen, however, their ardour towards her and their hostility towards one another increasing in exact proportion. Nor was there any decisive battle to end the bickerings between the brothers until on the afternoon of the fourth day after Burrilda's arrival, Wandana suddenly attacked.

The onslaught was as furious as the contest was one-sided. For a short while a confusion of cries was punctuated by the sounds of blows, most of them delivered by the powerful wings of the stronger and heavier Wandana; then silence.

A rain of falling drops spattered a trail through the reflections on the smooth surface of the water as the loser swept across the river just clearing the bank and zig-zagging wildly through the trees in a way of escape full of dangers for a flyer as large and as unversatile as a black swan.

With no rival to distract him, Wandana resumed his courtship. Mating followed almost at once and soon the pair were improving the site they had chosen for their nest in a reed bed, snapping off the reeds all around and heaping them in the centre of the space so formed. From time to time as the heap grew, Burrilda would clamber awkwardly with waving wings to the top of the untidy mass and trample it into some sort of firmness. Later, when she had laid her eggs, every living creature larger than a native mouse or a painted finch became their enemy.

On the day her young appeared Burrilda called and launched into the water; and they, no more than a few hours old, crowded after her. Often they would clamber upon their parents' backs and almost every evening the family returned to the nest in this fashion.

Summer wore on and the cygnets had grown into young swans when Wandana caught sight of the men. He watched them briefly, called briefly, and forged with powerful strokes across the river, his white wingtips flashing warning signals to other waterfowl. The rest of his family rose in his wake and the whole company swung round to pass high over the line of Aborigines toiling across the plain.

Wandana flew westward. He should have flown east, but shrank from the long flight over droughty country. So he continued to retreat until he came to the last of the big swamps. There was now only semi-desert before him with similarly waterless regions to the north and south and between him and the eastern coast.

However, the inland swamp still was a reassuring sight—a majestic expanse of blue, sparkling water dotted with islets and mud banks and beds of swaying reeds. The swans stayed mostly in the widest part and it was not until they paddled towards the edges that certain differences between this and the swamps of normal years became evident.

These differences were accentuated in the next few weeks as the shores drew slowly together. New beds of weeds were exposed, to be turned quickly into masses of decay; every day, too, the mud banks expanded and the many islets breaking the surface showed rims of green slime where swarms of insects hovered.

Except for one solitary pair of teal, all the water-fowl had left by the start of the nesting season; still the level of the water continued to fall, and still Wandana would not go. Sun-baked mudflats cracked into floors of brown mosaic, each mud tile curling like the petals of some unlovely flower; the water lilies died, and the supply of live roots and other aquatic growth dwindled. Then, suddenly, the teal were a-wing and they were such swift fliers that they were soon out of sight.

The swans were now alone in the midst of the desolation of a fierce drought. The sheet of water that had supported them so far was not only shrinking faster than before but was becoming stagnant as well, and it was extraordinary at this time that Wandana was not deserted

110

by his family, especially as Burrilda would often circle overhead as though about to fly eastward. Sometimes one or two of the younger birds fell into position behind her and, once the six of them formed the distinctive wedge-shaped flight of the wild swans. But still Wandana would not go and, always, the others flew back to rejoin him.

Time was running out, surely but by no means slowly, and, at last, faced with the certainty of starvation, the stubborn Wandana beat purposefully across the water and led his family eastward. They set out just as darkness was lightening into dawn, and before the sun came up to blaze once more upon the brown and dusty earth the company was far on its way.

Then, unaccountably, their progress ceased as the ever vigilant Wandana swung round in an arc, climbing on strenuous wings. The others kept their places and the whole company fought steadily upwards, crying to each other in the quaintly inadequate voices of their kind, all of them now aware of the reddish barrier of dust reaching to such a prodigious height that it filled the eastern sky. Even though the phenomenon was a great way off, it was also clear that a tremendous wind was whirling it towards the circling birds.

With their bills parted from the searing heat and dryness the swans fled back the way they had come. The altitude already gained was such that they were able to hurl themselves westward in a long incline so that the distance back was covered in less time than it had taken to fly east, and they planed down at last to the swamp they had left in the early morning.

The wind had gradually strengthened until small waves were curling upon the encrusted banks, tearing the fringe of green slime and churning up a greyish froth which was continually whipped away by the gusts. The afternoon darkened as the dust drew nearer, and the swans' frightened call-notes were lost in the harsh crackling of dead reeds in the wind and the slapping of waves upon the shore. Wandana worked towards a mudbank which had once been a submerged feeding ground but which now cut the swamp in two. With his tail feathers twirled and twisted by the wind he walked clumsily to its highest point, only to be blown brusquely down the other side and into the water again. Burrilda, and the others crowded behind him, hustled on by the violent gusts, calling to one another, clustering together, tossing in a brown fog of dust, craning their long necks as they peered alarmedly about them.

Half the sky was blotted out, and half the sky glowed with yellow

111

light which neither waxed nor waned until nightfall. By midnight the worst of the dust had passed, and the swans began to forage for food, though that was scarce enough in the unsavoury waters. However, much of the consternation and fear which had plagued them for so long had suddenly evaporated for, at last, Wandana had accepted with decision his responsibility as leader, the change in his demeanour being already plain to Burrilda and the young swans. His calmness calmed them. So natural is it for the wild things to interpret the behaviour patterns of one another that every member of the group knew a second and more determined attempt would soon be made to reach the coastal lagoons. But they were now content to leave the time and manner of that flight to Wandana.

Daylight showed the air to be much clearer. The wind had died. Westward, the red cloud was still visible, though far away; to the east the rising sun was a dull, copper-coloured orb.

The day was already burning hot when Wandana spread his wings, and dramatically every other wing also spread, beating as one. Long necks extended, webbed feet pattering over the bland surface of the swamp smashing its smoothness, the swans swept forward and rose into the air. As they gained height their initial efforts gave way to a rhythmic ease that carried them ever deeper into the heat of the day. In spite of its pollution the swamp had cooled, in some degree at any rate, every breath of air that drifted across it and it also had given up to the breeze some of its moisture. But the unhindered rays of the sun high above the earth now scorched the fliers and drove the strength from their bodies.

Wandana led, Burrilda flew to his right and just behind him, then the largest of the young ones, while the rest formed the longer arm of the V. A monotonous drabness broken here and there by clumps of trees, stretched to the horizon.

They flew on, without varying the regular stroke of their wings. They were ample wings—not pointed and narrow as the wings of the swifter fliers, but wide and somewhat rounded—and they might all have been directed by a single brain so evenly did they fly along their chosen way. The regular formation of the company, too, was maintained so exactly that no flier was ever out of position.

Nothing else travelled the sky, nothing moved on the plains except the dust clouds and the black swans seemed to be the only creatures left alive.

About midday, they slanted down to a waterhole, but there was no blue gleam to welcome them, no silver reflection. Wandana planed so low over the earth that he could feel a fiercer heat striking up at him. Curving between two clumps of trees on the bank he glided down more steeply only to check, then thrash his way furiously up, clearing the other bank and flying over the plains once more. The others too fled from that noisome pocket of slime and managed to lift their tired bodies again towards the unrelenting sun. The wedge-shaped flight kept travelling east.

That evening the earth seemed ablaze; and when the red reflection had faded and the luminous purple of night had taken its place there was still no glint of water anywhere in the world. Stars shone from the hot darkness, but the labouring wings did not falter, neither did they deviate from their course.

Very gradually, almost imperceptibly, Wandana began to lose height. It was easier to fly at a slight inclination towards the earth. And because the leader descended, so did his followers.

Long hours of exertion in the heat and lack of food were dulling his senses and it was not until the group swept low over a timbered ridge that he realized the imminence of disaster. With a further effort he checked this downward tendency and painfully gained height again.

The moon was almost immediately overhead when they alighted in a narrow gutter in an otherwise dry creek-bed—a gutter that reflected the moon as faithfully as a pool of good water but whose stagnant shallows contained no scrap of food. The thin leaves of an eucalypt whispered in the dry night wind and sometimes a puff of dust from the stricken plains drifted up like mist. Heat, moonlight, silence; a stirring of leaves and a whispering above, a wavering of reflections, then silence as before, the heat pressing down. The tired voyagers slept.

Wandana was eager to go in the early morning but his companions were listless from fatigue and daunted by the prospect of another day-long journey under a brilliant, pitiless sky where the air was like fire. So they remained clustered in a narrow band of shade under the highest bank, apparently determined only to stay—to die there at length from starvation and the heat.

But their leader's will to go on proved the stronger, and when the sun was some way past its zenith seven great birds with white wings seemed to rise out of the dusty earth itself, their shadows flickering amongst the broken channels of a dry watercourse.

The voyagers were not as strong as they had been the day before. Lack of food and the unrelenting heat had slowed their wingbeats. The

113

sun was low in the west when they reached the edge of the coastal hills. Driving his tired muscles to a more strenuous effort Wandana passed safely over the first crest of trees. Then his path sagged into the intervening valley. Burrilda maintained her place more easily, as did her offspring, except one. He managed to overtop the second ridge then flew down the line of the next slope, keeping to the same course. They heard him plunge through the branches and saw the last, despairing flash of his wings.

Wandana, too, was almost spent. Every stroke taxed his strength as his size and weight combined to drag him down. Gallantly, he fought his way up the next slope, his wings almost brushing the trees but, as he laboured over the crest, Burrilda called and, surging past him, took the lead. Drawing upon some reserve of power Wandana drove in her wake so that the following ridge glided under him unnoticed.

Burrilda flew on at the same height then, very gradually, climbed a little. Again the swans travelled in level flight; this time it was Wandana who led them higher.

It was dark when the swans went into a last, long glide, down to the lagoon where their leader had made his entrance into the world, and their arrival was so headlong and reckless a sousing into the cool water that it set up an indignant commotion amongst the ducks and the moorhens already safely floating between the wide shores.

After the newcomers had fed, they dabbled ecstatically in the shallows then rested on the open water, wings drooping from weariness. With the recollection of that prodigious journey through the furnace of drought-time already fading from his consciousness, Wandana drowsed as contentedly as any, caressed now by a wind from the ocean, damp and salt-laden; and by his side, the graceful, strong-winged Burrilda slept.

# The Hunters

BURRAMURRA, Ungalla and Cowarie, gaunt, black figures with spears in their hands, stood at the edge of a narrow beach and gazed over the grey waters of the Gulf of Carpentaria. Slow rivulets were trickling across the sand from the saturated earth, and it would rain again soon in spite of the brightness of a few blue patches between the piled-up thunder clouds.

The Aboriginal tribe was camped on a ridge immediately behind the beach—on the only high ground within a day's journey. North, south and west the sombre mangrove flats of Arnhem Land stretched into the steamy haze like another, darker sea.

The wet season had been unusually severe. Hunting was almost impossible, for the land had been turned into a morass and muddy rain water staining the shallows of the Gulf had driven the fish away. The women had long since failed to dig enough yams to keep a piccaninny alive; the lagoons where the lily roots grew were flooded out, and even the lizards and other small creatures had vanished. Burramurra's people were starving. Lately, they had scoured the Gulf close inshore but had met with small success. A turtle or two, some fish speared from the rocks, and a solitary dugong were all they had taken. True, the dugong had provided a feast for everyone, but the satisfaction had lasted only a few days; and many days were between them and the end of the rainy season.

Burramurra knew they must venture far out if they were to find another dugong. As the cleverest hunter he had been admitted to the councils of the old men who had seen such famines before, and they had told him that he must explore the waters of the Gulf at a great distance from land. And he had decided to do this because it was wise. He would take three canoes—one in charge of Ungalla, one in charge of Cowarie, with the big canoe for himself and his crew—and he would lead them out to sea in the late afternoon whether it rained, as it almost certainly would, or not, and whether the sea was rough or calm.

So, as daylight was beginning to fail, three dugout canoes were dragged to the water's edge by the lubras and boys of the tribe while the old men looked on, sometimes cocking an eye at the weather and muttering to one another. The canoes were surrounded by Burramurra's people, small and large, old and young. Most were excited at the preparations for the hunters' departure, but there were those who were already listless from hunger, and there were a few small children who just sat and stared, dull-eyed.

When he had made sure there were enough shells for baling, Burramurra took his spears and laid his hand on the side of the dugout as a signal to push off.

The three crudely-fashioned craft slid into the waves. The warriors at the paddles bent their backs, and the canoes crept towards the blank horizon. The uproar on the beach became fainter and soon the company of about thirty Aborigines was swallowed up in the steamy haze. There were only the slapping of waves against wood, the splash of paddles and the low, monotonous chanting of the men in Ungalla's canoe. The sea was not rough, and though small they were sturdy enough, these craft which had been hollowed from the trunks of trees.

Night came down quickly, and with it a deluge. Burramurra baled calmly. Ripples of light, pale streaks floating in the gloom, gleamed from the muscles of his companions. The paddles dipped and rose, dipped and rose, making faint luminous splashes in the dark. Every now and then Ungalla or Cowarie would call and Burramurra would answer. There were no stars, but no one questioned the direction of their voyaging. Scooping a handful of rain water, Burramurra drank, then pressed his stomach to keep away the pains of hunger. He flexed his muscles and called. Cowarie answered from the right and roughly abreast, Ungalla from the left and further back. Hour after hour they struck out into the Gulf.

There was a dank oppressiveness about the night which blanketed sound and brought the sweat in streams from the paddlers. They gasped for breath as they worked, and gradually pauses for rest became more frequent.

Burramurra was watching the east. For a while he and his men swung up and down, waiting. They could hear voices in Ungalla's canoe. Then, dead ahead, the star which belonged to the big reef on the other side of the Gulf shone between the clouds and the mariners knew their course was true. The three floating clusters of men waited until the moon rose and the sea around became a mist where the canoes showed as shapeless blurs.

The leader of the venture called warningly that they had arrived where they could expect to find dugong.

The low chanting of Ungalla's men died away; the splashing of the paddles ceased. As they stole on over the quiet, warm sea a growl of thunder vibrated in the humid heat, and Burramurra moved uneasily. He knew of the storms which swept into the Gulf from nowhere, and of the power of wind-gusts which struck like hammers and heaped the waves into rushing mountains of water. Then a dugout canoe was a small thing and the strength of a man was even smaller. But the thunder did not come closer and the tiny craft crept on more cautiously than before as though to escape its notice.

It seemed the old men of the tribe had been wrong or that some enemy was working against them, for they were a perilous distance from land and there was no sign of life in all that featureless waste.

Burramurra felt a touch on his hand.

Look!

The word was almost a breath.

Following the direction of the speaker's arm, he too saw what sharper eyes than his had found—an almost imperceptible glow of phosphorescence, a less dark line in the blackness of night, a sign which, were it pinpointed for civilised eyes, would yet have remained invisible.

The canoe nosed round, and detecting the change in the sound of the waves against its prow the other two crews ceased paddling. Burramurra rose to his feet, felt for his heaviest spear, and made sure that the line attached to it was coiled free to run. Taking up a firm stand he leaned forward, motionless as a carved figurehead. His men guarded their breathing, the paddles eased gently into the water and the rate of progress lessened until they were scarcely more than drifting.

The phosphorescent sign in the sea was plainer now, moving in the same direction as the hunters, but very slowly. Tense, with set jaws and lips compressed, Burramurra drew back his spear and held the pose. He could see the dugong's grey shape just beneath the surface. He heard the slap of its tail in the hollow of a wave and for one endless moment thought his prey had taken fright. But the shadow lazed on in the same phlegmatic, sleepy way, and Burramurra's spear began to quiver as he stretched his arm back for the throw.

He must not miss. Waiting until the quarry, a fat target, was wallowing almost under the canoe, he lunged down with all the force in his muscular frame. In that blow, too, was all the explosive energy of his pent-up excitement, and the thrust was so fierce that the spearman followed his spear headlong. Doubling expertly as an eel he emerged from the sea in time to clutch the canoe as it leaped forward. Deep-throated yells of triumph greeted Burramurra as he dragged himself aboard.

But the hunt was not finished, and though the heavy barb was securely planted it could not inflict a wound that would kill such a large creature. Repeated thrumming jerks on the rope told of the frenzy of the captive fighting its battle against the cleverest of hunters.

Burramurra turned to his companions and laughed, his white teeth gleaming in the light of dawn. At the same moment the bow swung, and held down by the dugong's powerful, diving turn, nosed into a steep-sided wave. Water poured inboard and a tiny billow rolled swiftly aft, splashing past the paddlers and washing the baler shells overboard.

Pausing only to exhort the others to hold to their prey at all costs Burramurra soused into the sea. When he returned to the surface he had a shell in each hand and the canoe was moving away from him faster than he could swim.

Philosophically, he began to tread water. There was no breeze and the swell was unbroken by any fleck of foam. Overhead, the sky was as grey and as heavy as it had ever been. Ungalla's and Cowarie's boats had disappeared but he could see his own canoe still moving steadily away. He waited, knowing the others would return for him no matter how far the dugong swam. He could see the paddles flashing as his men fought to bring their craft round but it passed almost out of sight before it turned and all the while the man in the water swam leisurely in pursuit, firmly clutching the two precious bailer shells.

When the canoe at last headed directly for him Burramurra knew that the dugong must be nearly spent, for a warrior in the bow was steering it by tugging on the rope. A few more strokes took the swimmer into the path of the canoe. He clutched its side, handed up the shells, hooked his skinny black leg over and swung himself out of the water. Once again in his position in the bow, he gently hauled in the rope until the quarry's tail was flapping weakly against the wood, when he passed the rope to the man behind him and picked up his smooth-bladed killing spear.

It was the moment of victory. Superbly balanced in that tossing primitive craft, with his spear poised and every muscle of his wet black body agleam in the wan dawn, Burramurra drew back his arm and, to the shouts of his companions, plunged the blade deep into the dugong's seamed back.

There was a wild, white flurry in the sea, another outburst of shouting and then suddenly silence.

In the early, cloud-obstructed light the five warriors, now slack-jawed from exhaustion, stared down at the carcase floating with its leathery sides awash. Wisps of blood trailed from the spear wound and the creature's mouth set low in the blunt, cowlike muzzle gaped wide open. They discussed the prize, pointing out its extreme fatness, its girth and its length which, Burramurra said, was equal to that of two men—which was an exaggeration. He ran his small, nervous hand over the brownish hide ripped by the sharp coral of the reefs, and pressed his fingers into the layered flesh.

'Fatter than old Darganna,' he chuckled, but the others looked away, for it is a bad thing to speak slightingly of men who are so full of wisdom that they know how to make yams plentiful and wallabies to multiply and where to find the biggest dugongs when the shallows are murky with rain running from the land.

Leaning over, Burramurra seized the shaft of his embedded killing spear and setting his foot against the carcase, pulled out the scarlet blade. Nor was the labour of securing this mountain of food yet completed. It was out of the question that they should try to tow it back as the drag of its bulk would be enormous and also the taint of blood would attract sharks. It was too heavy to lift aboard, so the only course left was to submerge the canoe and manoeuvre it underneath the prize.

Black heads bobbed in the grey water and a line of arms pulled the canoe over to let in the sea. They tilted it enough to float it under the dugong. Relieved of the men's combined weight both wooden sides of the craft reappeared. Burramurra crooked his elbow and fished up the shells from the bottom of the canoe. Handing one to a companion he began to bail furiously, grunting encouragement to the others as they strained to widen that vital margin which turned back the sea.

The swimmers rested; then the shells again were plunged into the mess of blood and water swashing round the dugong, and as one red gout after another splashed into the sea the craft began to ride more buoyantly. A warrior climbed over the dugong and straddled it, leaning forward and dipping the bailer shell in front of its snout. There was not much room to work in, for the victim was almost as long as the canoe and where its girth was greatest it swelled above the sides.

The hunters were a quaint sight as they set out on the homeward journey. Three of them were obliged to straddle the prize and had to reach out awkwardly to get their paddles to the water.

If anything the morning was more oppressive than the night had been. Heavy showers beat the waves to an oily smoothness and bailing continued almost without pause. Then, at midday, the rain ceased and the sun was sailing in the clear. Sweat dripped from the labouring warriors and the sunlight played across their skin like flame. The sea burned blue. Sharp arrows of brightness glanced off its glassy surface to strike up at the paddlers fiercely. Shifting his position on the monster, Burramurra shaded his eyes and stared hopefully ahead. There was no land. The paddles rose and fell, stroke for stroke, their wet blades flashing like steel. Red pools in the bottom of the canoe ran forward and back, forward and back, ceaselessly; the dugong's tail lolled from side to side with the motion of the waves, and the musty stench of the kill became stronger.

Burramurra took the paddle from the man next to him and bent his back. No one spoke. Wearily, mechanically, with cast-down eyes and drooping eyelids, they drove on through the heat.

'I see land,' a tired voice muttered. The speaker's companions did not bother to look.

There were hours of paddling between them and the blurred line showing above the horizon. Distance was an old enemy of theirs. But when they saw black dots moving on the beach their strokes quickened. The other two canoes had returned long since and were lying like curled chips on the sand.

While still out of earshot Burramurra let out a shout of triumph and roared that the dugong was as long as an old man crocodile and as fat as a piccaninny. He began to sing and the others joined in, keeping the rhythm of the stroke.

As the canoe hissed on to the sand it was surrounded by a frenzied, gleeful company, all anxious to prod the carcase with a testing finger and to congratulate the hunters.

They rolled the dugong from its cradle. It was bigger than the one which had fallen to Ungalla's men on their last successful expedition two long hungry months ago and many were the compliments lavished on that unlovely mass of flesh.

During the feast that followed, Burramurra recounted in detail and with some embellishment, the story of how he and his crew had surprised, speared and worn down their prey; and once or twice Ungalla murmured something about his own stale triumph. But Cowarie and his men said nothing since they had nothing to say.

# Long Reef

SWEEPING over Long Reef the downpour blotted out the houses on the coast road. The peninsula linking the reef with the mainland, enclosed by curtains of rain, might have been 1000 miles from civilisation instead of on Sydney's front doorstep. The peninsula's smooth, treeless slopes blurred to the grey of a gull's wing, and the noise of the ocean softened to a peaceful monotone.

But seaward, on the exposed rocky plain lying beyond the cliff's edge, the clamour of wind and wave merged into one continuous roar. White crests, racing out of the dimness, arched themselves menacingly above the reef, and at every moment a towering mass of water threatened to engulf it, only to burst and spend its strength among a ring of guardian boulders.

A flight of terns whirled up through the clouds of spray and swung screaming down the wind. They turned and balanced awhile in the gusts, then beating back under the lee of the oncoming rollers, again settled in a disconsolate group. And, as the last of them alighted, there came shrilly a clear, far-carrying piping.

The calls grew more distinct, there was a glimpse of many wings, and a flock of small grey birds wheeled down to the reef. Instantly that part of the rock was transformed. Racing over its weed-covered surface they skirted the pools the tide had left, explored the shallows, took short, quick flights across their hunting grounds, and weaved such intricate and swift-changing patterns that the eye soon failed to see more than a swirl of grey motes against a darker background.

As the tide flowed deeper through the boulders the newcomers were driven back towards the cliff. Waves began to spill over the outer rocks and the birds rose, flashing inland with the erratic flight of all waders; for, though not much larger than sparrows (and though burdened with the name of little stints), these tiny creatures who had just travelled thousands of miles to escape the northern winter were still strong of wing.

They had nested in Siberia—perhaps even in the far north of Siberia—and, though their route had lain mostly overland, it would not be reasonable to assume that such a journey had been merely a succession of easy stages. If they had followed the coastline as much as possible they had also been faced with many long flights across the ocean; if they had paused to rest in sheltered bays, it is also certain that they had encountered gales and rainstorms in the course of their extraordinary migration.

Long Reef would be their home for the next few months of the year. At high tide it was visited by the cormorants, who basked on the exposed knobs of rock or fished in the green-tinted water, but as soon as the tide ebbed the stints returned from the margins of a nearby lagoon. Always they would be joined by the red-capped dotterels from the beach across the bay, and the whole company would forage amicably together. So they lived—the dotterels unobtrusively, the stints dividing their time between the lagoon and the rocks, feeding, wading, flying, and presumably sometimes sleeping, though they seemed never to be still.

It was spring when the curlews arrived. They flew low and straight, about ten feet above the wave tops, and came to rest on the extreme point of the reef. There they stayed, watching with bright, suspicious eyes. Sometimes one would jab experimentally at the rock with his long, curved bill, and at intervals their mournful calls carried far inland.

The following morning they were gone. That afternoon, perhaps, they would complete a half-circle of the earth by planing down to some secluded mudflat.

The popularity of the reef grew. Its weed-covered expanse teemed with marine insects and crustaceans. Some of the migratory birds made it the terminus of their journey, but compared with those passing over their numbers were small. Only the gulls, terns and, to a lesser degree, the redbills regarded it as a dwelling place.

The redbills, from a distance for all the world like clumsy crows, improved upon closer acquaintance. They were almost black, with bright red bills and long red legs—intent, conscientious feeders who poked and pried among the weed while the tide was still running out and from then until the water again became too deep. Each took the greatest interest in his companions' welfare, and whenever one found troublesome prey the others would come flapping and bouncing and squawking towards him, remaining until the food was won or filched by the sharpest.

Once when the tide was high a pair of golden plovers flashed overhead and, wheeling, settled where the red-capped dotterels were racing along a nearby beach above the hissing curves of the waves. Each was about the size of a dove, but longer of leg and swifter and stronger in flight. At rest, they blended so exactly with the colour of the sand that they were almost invisible. Their journey, too, had begun somewhere in the tundras of Siberia, and though they had not come far that day, they rested until the water receded.

When the tide was conveniently low the flock of little stints returned to the reef. The dotterels followed, then the plovers. A crisp breeze whipped the tips from the waves and blew the waders' plumage into ragged tufts whenever they turned. The flash from every reflecting surface burned; yet, though the birds were thick as sparrows round a handful of wheat, they were hard to distinguish even from a short distance, for the flicker of their swift movements and the glint of pale breasts blended confusingly with the ripples on the pools, the dart of wind ruffs across the water, and the white flare of foam on the outside rocks.

In contrast to the others, the golden plovers were rotund and staid. Often they would stand as though in profound meditation while the smaller birds darted earnestly about them.

Other feathered travellers, endless flocks of them, were winging south, and with each group there would be a quick interchange of calls before the fliers faded into the distance. Stragglers were constantly descending to rest, then hurrying on, but there was one more migrant to make the reef his journey's end. He was a grey sandpiper, and he stayed by himself on the streaming outer rocks—a slim, grey bird with quick, untiring legs and a long, pointed bill. He moved jerkily but swiftly, almost under the curve of the waves, and if he was more wary of the dangers that might come from landward he was more contemptuous of the dangers of the sea. Often he would wait until it seemed a speeding line of foam must dash him under, then at the last instant he would rise with a quick flutter of wings, swing sharply, and drop back to the rock. And his high-pitched, double whistle was clear and shrill, and the flash of his wings was sure.

Living where the spray rained upon him and the white foam swirled incessantly, the grey sandpiper blended with his background of tumbling water, and the quickness of his movements was in tune with it.

He was there long after the rising tide had driven the others away.

With the first heat of summer, flocks of double-banded dotterels began to drift out from the salt mudflats and river margins.

A few joined their red-capped kinsmen and imparted some of their restlessness to these stay-at-homes. The double-banded dotterels took high, long flights. Already their breasts showed the black and chestnut bands of their breeding plumage, and the urge to travel east was strong. At last they spiralled up, as they had done so often before, then, no longer zig-zagging, made out to sea. Far ahead a high-sailing blur swung in a circle, the latecomers flashed into its midst, and the travellers pressed on.

Their passage across the Tasman Sea was perhaps the most wonderful of all, for there are no resting places on the way to New Zealand.

When the dotterels had gone the birds on Long Reef settled down contentedly once more. The summer days with their blue banners marched on, and by the middle of February small parties of outward-moving birds had swelled the flock of stints to twice its original size. They began to take short flights over the ocean as the double-banded dotterels had done, and even when feeding on the rocks they were restless, flying at the slightest excuse—a noise, a distant movement, a passing cormorant or gull. Anything was sufficient to disturb them.

Suddenly, they were heading for the vast Siberian tundras they had left only six months earlier, and the great tide of bird migration,

126

now returning north, gained impetus and volume with every week. Each night, high up against the stars, the winged hosts passed. Sometimes a faint flickering across the moon betrayed them, and, incessantly, far away, clear calls came piping down the wind.

The breasts of the golden plovers turned to black, and finally they stood in full breeding plumage—brown and gold-flecked backs, a white stripe over-arching each eye.

Soon they too were drawn up by the streaming voyagers, and only the gulls, the terns, and the little red-capped dotterels remained on the reef—they and the grey sandpiper. Foraging unconcernedly along the foaming edge of the rock, this last, alone of the migratory birds, seemed indifferent to the passing armies.

But, in due time, the call reached him too. It was dawn, and the ocean was like a rippling sheet of grey silk. At a barely audible whistle the sandpiper looked up. The red arc of the sun lit a rivulet of wings spilling down the sky, and, rising, the last of the migrants mounted into the brightness.

Long Reef stretched into the sea, its weed-covered flats deserted. But the tide would flow again and, as surely, the birds would return.

# King of the Skies

NIGHT ebbed slowly from the peaks of the ranges and flowed down the valleys and over the edge of the world; then the sun's first rays touched the towering crest of the rock and lit with sudden clarity a forest of twisted trees swaying and trembling in the wind.

Cloud shadows came to fleck the green waves of bush and as the full brilliance of day flooded across the vastness the western slopes of the ridges darkened. Meandering between the hills a narrow water-course glinted like a broken silver thread, and constantly across the whole wide panorama piped the grave melody of the wind.

The rock, the dominant feature of the scene, was not only a gigantic monolith in its own right but stood on the highest point of the range. It was inevitable, therefore, that the trees growing in a pocket of earth on its lofty summit had always been exposed and would continue to be exposed throughout their lives to the full force of every storm that blew; and to this their stunted and mis-shapen limbs bore witness.

Restlessly—he might have been the very spirit of that wild place— a wedge-tailed eagle spread his great wings and lifted buoyantly above the confusion of swaying branches, only to drift down again and alight on his former perch.

Boldly drawn against the sky he watched a winnowing hawk pat-rolling his hunting grounds far below and marked, away to the east, the effortless soaring of another of his kind.

Calling loudly—a shrill, clamant cry—he again arched his long wings. There was a savage grace in the movement, just as there was power in the width of the great bird's back and in the disproportionate size of his muscled legs and heavy, needle-pointed talons. Fierceness, too, was in his narrow head and hooked beak, and in the piercing stare of his deep-set eyes—brazen and unwinking, miracles of sight, golden brown, unfathomable.

With sudden impatience he called a second time to his mate still drowsing in the draughty dimness of their eyrie, then beat above the trees and swept over the abyss in a magnificent downward curve to swirl close to the rock, one sombre pinion almost brushing its wall, finally to veer away again, setting his course west.

Unhurried, with wings unmoving except for delicate readjustments to the airstream, he rose in wide, swinging circles until he was soaring at a plane that foreshortened even the rock.

Far below a moving speck curved slowly away from the eyrie and a thin, far-away call came faintly to the first flyer. He headed into the wind, balancing and see-sawing gently.

The speck eddied steadily aloft, the male wedge-tailed eagle waiting until his partner had ranged alongside, then in smooth rotation, one moving clockwise, the other counter-clockwise, they passed and repassed, always drifting west, each flowing revolution overlying the path of the last.

Water in pools and shrunken streams continued to gleam occasionally through the swimming deeps of sunlight but the wide forestlands which had been so richly green were becoming gradually browner towards the foothills of the ranges. The plains were drier still, the shadows of the eagles sliding smoothly over dusty earth. Further west, the level country was veined with dry creeks and chequered with red clay pans.

Viewed from that height the world moved in slow motion. A puff of dust whipped up by the wind dissolved gradually into nothingness; a flock of crested pigeons rocketing from one feeding ground to another, for all their swiftness of flight, seemed to move at snail's pace.

The eagles' tawny eyes, shaded by ridges of overarching feathers, caught sight of a mob of white cockatoos, saw them raise their sulphur-yellow crests and screech mutely. But in that lonely ocean of space where the streaming currents swirled unhindered, no sound reached the birds of prey except the hiss of the airstream sliding from their feathers.

Soaring on an updraught, the larger of the pair poised with the gale droning through her pinions, swung away and raced downwind, turned again as a speeding wave of air shrilled under her muscled wings and then dipped lower in a long, leisurely curve.

The fliers had now left the mountains far behind. As they travelled

130

steadily west patches of grass sometimes appeared, where perhaps there had been a rainstorm some months before; but generally there was only the grey and brown and red earth of drought time—a land of ochre colours dotted with the dark bronze-green of infrequent clumps of trees and scribbled over with dry channels that had lost their way.

Simultaneously both eagles dived, and checked. They hung there, the male bird entirely motionless, his consort teetering and rocking a little, her wing-tips making small correcting movements as though she were caught in an air turbulence so slight that it was scarcely worth the effort of rising or descending to a calmer place.

A mob of kangaroos was moving slowly, very slowly, towards the ranges where there was at least some feed. They paused often to nose in the dust for dry grass stems. They had come in from the back country where the drought had locked down in an iron grip and they were weak from lack of food and water. But they were full-grown animals and any thought of attack would never have occurred to the eagles, except, perhaps, if a particular member of the mob had been too weak to keep up, too weak, almost, to move. But such was not the case in this instance. However, there was one laggard and it was he who had attracted the attention of the hunters. They descended further, and again checked, to hover over a thicket of wilgas where a small, dark form was making its way through the maze of dense-leaved trees.

Though less than half grown the kangaroo seemed to be under the delusion that he could not possibly have an enemy in all the world. Already half a mile or more behind his companions he showed no inclination to rejoin them and it was some time before he set out in leisurely pursuit.

The eagles watched, no longer sailing but hanging in the air, gradually descending. Immediately below them, the young kangaroo, moving unsuspectingly on his way, was skirting one of the many dry watercourses in the vicinity when the smaller of the eagles struck. Timing was perfect, the power of the blow hurling the quarry so violently forward that the attack by the second member of the pair, following close in her partner's wake, completely missed its mark.

But the respite was brief. As the kangaroo bounced back to an upright position he was raked again by the male bird's talons, then hit by the heavier, speedier female.

If their prospective prey had been even slightly less sturdy than in

131

fact he was then the kill would have been made swiftly, but the young-ling had not yet been weakened by the drought; and, also, by luck or by instinct, he had headed back towards the wilga trees which were much closer than the main company still dawdling towards the hills. Even so, the eagles almost succeeded in preventing the terrified fugit-ive from reaching safety, knocking him over time after time with talons driven by the full weight of their plunging bodies. Sometimes the impact sent the dazed little creature slithering headlong through the dust, sometimes it merely spun him momentarily offcourse; but always the deadly claws raked him. Though a good deal of fur had been ripped from his back and though bleeding from many a slash, yet he remained as resilient as a dingo in recovering his balance and simply kept press-ing towards the safety of the wilgas.

Occasionally, the eagles screamed angrily but for the most part the only sounds were the thud of blows, the swish of wings and the muffled thumps of the kangaroo's hind paws desperately thrusting him for-ward. Finally, he gained the trees. Another minute or two of that twin assault and he would have been spent; as it was, the eagles swung away and circled high above the dangerous branches.

The kangaroo sat motionless in the shadows. For a while the eagles continued to wheel overhead then they rose effortlessly, as though swept skyward in a violent spout of air.

The crow seemed to materialise out of the steely blue light of the noonday sky. He was flying high, though not as high as the eagles, and his course was straight. Sometimes he called, and his cry might have been the cry of the drought land—harsh—despairing—in its long-drawn trailing-off into silence.

Wings beating with rhythmic slowness, the traveller descended towards an open forest of eucalypts and alighted on a dead limb that shone like silver in the sun. For a while he searched the plains all around, then watchfully scanned the dome of sky. He croaked softly, his glossy throat swelling, dropped to a lower branch and peered fix-edly down. Perhaps the wavering leaf-shadows gave the sheep some appearance of life, but whatever the reason the bird was in no hurry to approach. Mockingly, he made obeisance to the dead then waited, uttering long-drawn sighs of sympathy and murmuring and posturing in grotesque indecision.

Each time he moved the sunlight glistened on his sable plumage and momentarily contrasted his sleekness with the faded world about him.

132

Suddenly he called with all his power—a discordant scream that towered into the hot skies and ended in a regretful moaning. He moved to another tree and looked up to mark the lofty flight of an eagle before again turning his attention to the matter in hand. He considered the sheep from his new position.

A willy-willy spun past the edge of the trees and the crow grumbled huskily and planed to a pile of fallen branches.

Again he hesitated, apparently on the point of retreat, then hopped on to the carcase, pecked once, and flapped hurriedly upward with a shred of wool trailing from his bill.

While he rubbed the wool free, two of his kinsmen swung down from nowhere. The newcomers were followed by others and there began a metallic whispering, a low, unmusical chorus with frequent silences. Dark shapes floated between the tree trunks, almost brushing the sheep as they passed.

A crow looped and settled, there was a quick rustle of falling wings and, in a flash, the carcase was hidden beneath a vociferous, quarrelsome cluster of birds, all tugging and wrenching with sinewy strength. Their former wariness vanished, the hushed expectancy of a moment earlier gave place to shrill complaint and scrambling activity, and so intent were they upon the feast that a great shadow swept over them unseen.

The wedge-tailed eagle wheeled in a driving arc and hung in a rising air-current.

From the confusing patterns of light and shade a score of black wings flicked back the sun.

The eagle fell a thousand feet; his consort also came down in one spectacular dive and balanced alongside him.

As the two birds of prey dropped into their midst, the crows rose like a whirl of charred paper in a gale, some circling above the trees, others perching among the branches, and all maintaining a deafening clamour of indignation.

Indifferent to the uproar the eagles landed clumsily and partly hopped, partly shuffled towards the sheep.

A crow poised overhead, side-slipped, and settled a few feet away. He edged in knowingly, head on one side, then reached out. An eagle jabbed at him and he skipped aside with a croak of alarm.

Other crows ringed themselves about the feast and the circle became smaller. They seemed more like reptiles than birds as they extended their necks at the stronger robbers and inclined toward them, bobbing and hissing viciously.

133

Sometimes one would run swiftly in and filch an unguarded fragment, only to be caught by his companions. At first these intruders were met by the dart of a hooked beak or the flick of a mighty wing but, as the pilferers became more numerous, the eagles tired of driving them back.

Like a crowd of disreputable, black-coated beggars they skipped in and out, seizing a morsel here and there, tugging at tough shreds or pursuing some companion who was fortunate, or unfortunate, enough to be in possession of booty which could not be swallowed at a gulp.

In the midst of this restless, bickering mob the eagles now fed with the careless air of the powerful, though at intervals they would raise their cruel heads and stare into the shadows of the trees.

Suddenly, cutting through the quarrelsome undertones came the snap of a stick, a quick rattle of leaves and, a moment later most of the crows were in the trees while those still on the ground were tense with watchfulness.

The eagles arched their necks threateningly but the noise continued—almost comical in its loudness—as though a boy were kicking up leaves and branches as he walked.

Then the sound ceased and the flat, triangular head, armoured and grotesque, which had thrust over a fallen tree trunk watched with eyes that were black and sharp and unwinking. When the perentie moved again it was with startling directness. He clawed his way over the trunk, flopped heavily to the ground and calmly, insolently, wad-

dled closer. Each step seemed a labour for the weight of his long body was great.

The eagles yelped angrily and the hardiest of the crows flew up to join their kinsmen in the trees.

The aplomb of the newcomer was superb. The most threatening gesture failed to impress him, failed even to arouse his curiosity. He slithered down a shallow, crumbling bank, blundered through a litter of debris, then, scarcely out of reach of the eagles' beaks, stopped. It seemed incredible that he had ever moved. The eagles stirred uneasily, for the perentie combined the menace of the crouching beast with the inertness of a fallen branch.

Suddenly, a wing-tip slapped at the moveless snout and, in the same split second, the perentie surged forward.

There came the clash of armed jaws, a thrashing of wings, and in place of the eagles the big reptile was now straddling the sheep, tearing at it with convulsive lifting wrenches of his powerful neck or chewing with his head turned sideways in the manner of a hungry dog. He paused to force a gobbet of flesh down his throat, then backed away, jerking and worrying at a tough strip of meat.

Again the crows descended but this time did not approach. The perentie's sideways glance checked each would-be visitor; the efficiency of those industrious jaws was a danger signal.

An alarmed caw was the only warning of the eagle's attack.

He hissed through a gap in the dangerous trees but swung up as the

giant goanna whirled with jaws agape. The bird re-joined his companion and the pair spired and circled, then with high-pitched, whistling cries of rage dived again. One came hurtling down and looped, the other swerved as his enemy spun to meet him, an outstretched wing catching a branch with just enough force to tug the flier off-course. Immediately, he lost height. Nor was there much chance of recovery in that confined space and the lower the eagle sank the more he was balked and harassed by other branches and clusters of foliage so that, for all the noisy flappings of his wide, dark wings, he arrived at last with a final jolt upon the ground.

Ruffled of plumage, shaken and somewhat alarmed the great bird peered between the tree trunks, then started to shuffle awkwardly away in an effort to rise again. The perentie followed, paused, then, amazingly, seemed to turn into another creature as he reared into an upright position and gave chase at a speed and in a manner extraordinary for an animal of such length and weight. For he was balanced now on his hind legs, running as a biped, tail clear of the ground, front legs hanging limply. He seemed to have doubled in size.

The eagle was already flying, but very slowly because of the many obstacles in that alien environment; unassisted, too, by the still air near the ground. Desperately, the labouring wings beat on, almost brushing the earth at every stroke.

The overtaking of the quarry was an easy matter and to bring the eagle down was not difficult; but to subdue it was a stern task even for an assailant as strong and as tenacious as the perentie, now almost hidden in a turmoil of wings and a thickening cloud of dust. Then, abruptly, stillness. But not silence, for the bird's cries, sometimes harsh, sometimes shrill, but always terrified, were as loud as ever.

Eventually, however, even that last sign of resistance died away and the eagle's head drooped. The hooked beak stayed agape. The loser was gasping now, sprawled out, exhausted. The perentie, his grip as terrible as ever, crushed down with his full weight.

The larger of the two eagles, flying agitatedly and erratically above the trees, was now presented with a fair target. Hovering a moment she dived with wings almost closed. There was the thud of impact and carried on by her own momentum, she rose in a long curve and beat up through a gap in the trees, ready for a second attack.

But no other assault was necessary. Hurled clear of its prey the perentie showed no disposition to dispute the matter further but froze into absolute immobility, concerned only with defending himself against the dangerous sky.

136

The victim of the giant lizard's unexpected speed as a pursuer and skill as a fighter took some little time to recover enough strength to make a second bid for escape, that dazed and battered eagle eventually lumbering away through the shadows to flap unsteadily into the air and then, finding a rising air current, to lift more confidently above the trees and rejoin his partner. The two were soon out of sight.

For perhaps another five minutes the perentie showed no sign of life except for a little blood welling from cuts made by the eagle's talons; then abruptly his head turned as though snapped around by a spring, the long, forked tongue flickered and he seemed to become aware again of the activity and noise within the small forest. His eyes sought out the carcase of the sheep though, in fact, it was almost hidden beneath a horde of black-plumaged crows, their waving wings flashing back the light.

The place was full of sound and movement, of rustlings and croakings and scufflings as the birds bickered amongst themselves for better positions and a larger share of the booty. There was a murmurous hush, however, a sudden quieting, as the strongest contender of them all, the effect of the attack which had so surprised him now worn off, once more advanced to claim the spoils.

As before the crows flew into the lower branches of the trees. But they descended again, singly, then in twos and threes, and in groups until the ground nearby was covered. But the closest were careful to keep their distance and if any ventured too boldly there would soon come the warning snap of armed jaws.

But at long last the perentie was fully fed and after a few ungainly steps, came gradually—in slow motion, as it were—to a stop; and there he seemed content to drowse.

The crows sidled nearer, fearful even yet. For the strongest claimant was still close enough to the dead sheep to be able to reach it in a single lunge; and at that range—something less than his own length— he was as quick as a striking snake. More crows were planing down from the surrounding trees, seeming to materialise out of the sparse foliage. In tones modulated to match their funeral garb they discussed the question of safety, whispered to each other with sepulchral voices, confidentially, encouragingly. By degrees, the gloomy circle contracted and the mourners drew in.

Still they hesitated, then one called harshly, longingly, and with rigid outstretched neck approached the corpse.

# Reprieve

WHEN the alarm clock rang the only evidence of approaching day was a faint silvering of the window frame. But the boy immediately swung his legs over the side of the bed and sat up, yawning and running his fingers sleepily through his hair. Having arranged his clothes in their correct order the night before he was quickly dressed. He slipped past the other two bedrooms and into the kitchen where his shoes made no sound on the earth floor.

Steering by the grey-glinting beacons of a few well-polished pots hanging on the walls and the bright knobs of the fuel stove he cut a thick slice of corned beef which he sandwiched between two slabs of bread and crammed into the pocket of his jeans, to be eaten later. Then, having replaced the lid of the earthenware crock to the sound of that deep, bell-like note which always seemed to occur no matter how careful he tried to be, he lifted a pair of field glasses from their special nail behind the door and stepped outside.

One or two stars were still to be seen in the water-white sky. There was no breeze; the air was cool, and the morning quiet in spite of the distant sound of the ocean. Beyond a huddle of outhouses, themselves indistinct in the dimness, loomed other, vaguer shadows—shadows the boy could have put a name and shape to, he knew them so well, but shadows that were mysterious too—at least whilst they were awaiting full recognition by a new day.

He moved quickly across the small garden, eager to get away from the house. For the house never changed. It was the outside that changed, and after an absence of nearly a year he had arrived home only yesterday evening, late, long after dark.

Lines of bright but colourless dewdrops flashed down from the fence wires as he closed the homestead gate. He kept to the spine of the ridge which was also the highest part of his father's farm.

The ocean lay about a quarter of a mile distant and the sound of it became louder as he drew further away from the house and its protecting trees. There must have been storms out to sea for the rollers had been crashing on the shore all through the windless night. As the light strengthened he discovered that the sand dunes stretching north and south were completely hidden in a fog of sea spray.

From anywhere along the top of the ridge it would have been possible ordinarily to look over the dunes and see the beach and the edge of the surf. To-day it was not.

To the west, however, the air was clearer and though there was a thin land mist, he could easily discern the top of the escarpment that thrust like a wedge into the paddocks of his father's cattle-breeding property. Sometimes the boy thought of the escarpment as the prow of a huge ship furrowing green waves of grass.

He slowed his pace. The ridge started to descend a short way ahead and he wished to make a general survey from that vantage point before visiting certain of those special places of interest he had in mind. A diffused red glare tinting the wall of spray and mist to the east showed that the sun was already above the horizon. So he seated himself on the top rail of a fence, disposing of his rather meagre breakfast while he waited. He heard the unmistakable call of a whistling eagle but could see no sign of the bird. It was probably circling over the estuary—a favourite hunting ground—and would almost certainly be flying at a great height.

A breeze off the sea, growing stronger, brought with it the cries of a flock of gulls; nearer at hand, a magpie gave its fluting call and the boy glimpsed it volplaning past.

Under the combined influence of the breeze and the rising sun, however, the mist was rapidly lifting, first from the shoreline where it had been thickest and then from the grassy slopes and hills—lingering awhile along the creek and under the escarpment.

Though they had been dull grey not long ago the dunes had now turned a pale golden colour, patched here and there with yellow-green grass and low scrub with dark foliage that in the distance seemed black. The boy raised his glasses to enjoy the spectacle of a leaping white surf, its rising spray being dispersed by the breeze.

A flock of gulls, flashing in the sun, floated into the glasses' field of vision, rising and falling as they played with the notion of making headway. Sinking slowly down they vanished behind the dunes. There would be gannets diving further out, and possibly shearwaters. Two crows came across the paddocks, seemingly in danger every moment of brushing the grass as their deliberate, black wings thrust them steadily on. They travelled below the breeze for much of the way until they too alighted on the beach. The boy could imagine them walking just above the reach of the spent wave, sometimes working their way further up the slope of powdery, wind-rippled sand, searching for the dead things of last night, or last month—swaggering unhurriedly along, their heavy bodies swinging awkwardly with every step, their sheeny, black plumage glinting like metal in the sun.

The wide strip of pastureland between dunes and escarpment was streaked haphazardly with yellow sunlight and geometrically divided into rectangles by brown post-and-rail fences; and now the boy could scan all of his territory. He did not see it as being bounded by fences, but rather separated into zones by the natural features around him. To the north, behind the house he had just left, there was an expanse of low sandy undulations covered with bracken fern which had always seemed to him to be an extension of the dunes bordering the beach; to the south, of course, was the wide estuary and then there were the shoreline and the sea. The western escarpment standing out boldly, the sun on its towering cliff face, was a barrier guarding a rugged hinterland—a great promontory of rock with a crest of trees in black filigree against the sky.

Heaped against the cliff's base were the accumulations of detritus that had fallen down over the ages, and below these slopes again a patch of the original rainforest that had once flourished over big areas of the district.

The boy loved the pasturelands, but for him they held no secret. On the other hand there was always mystery in the untouched dunes, the shape of the estuary, unchanged for perhaps a 1000 years, and the remnant of rainforest. He knew the forest so well, especially its floor of shattered rocks, its great trees and strong lianas. Yet it, too, would have been grassland, he realised, except for the ancient boulders that had tumbled from the escarpment over aeons of time, a torrential overflow from the heaps of detritus piled at the foot of the cliffs, a mass of broken rock gradually bound together by vines and tortured tree trunks and questing, spreading roots, the whole cemented by an unceasing, sedimentary downpour of leaves and twigs and dust.

'I'd soon have it knocked down if it would give me more grazing land' the boy had once heard his father explaining to a neighbour. 'But

141

where's the point in clearing country when all you're going to get out of it is an expanse of Brobdingnagian rubble? Can you tell me that?'

Evidently the neighbour could not. He was not too sure of the meaning of Brobdingnagian but he did agree that the thought of allowing valuable beasts to wander through a litter of gigantic boulders was too ludicrous to be entertained for a moment, even if a few miserable blades of grass had managed to waver their tenuous way into the open.

Almost everywhere the boy looked he was able to add many details to those the field glasses had power to reveal. Certainly, the glasses showed him the external wall of the rainforest with great clarity, but it was his mind's eye that took him into the gloom of the forest itself with its confusion of wet and mossy boulders and its labyrinth of growing things; it was the field glasses, too, that transported him to a fixed point close to an islet in the creek where a pair of wood-teal had nested every year, but he would have to be much closer and be able to move about cautiously, this way and that, if he were to have a chance of finding the sitting bird, concealed as it undoubtedly would be by moving shadows and waving grass stems. It would be watching him—not yet, but later, as he walked down the hill to the creek.

Without the help of the field glasses, however, he would never have glimpsed the swift zig-zag flight of a small flock of waders—probably snipe, though they were too far off to be positively identified. They alighted on the edge of a swamp where the creek's flow, interrupted by a series of rock ledges, spread out widely on either side before draining away into the salt marshes and tidal flats of the estuary. He watched awhile but the waders did not take to the air again, so he went on down the slope towards the creek.

Soon the breeze was cut off by the ridge between him and the sea—a ridge that declined as it neared the level country around the estuary. From close range the boy turned his glasses on the islet and soon discovered the wood-teal, neck craned alertly, eyeing him through the grass stems.

A short distance further on he put up the same flock of waders he had seen from the ridge and followed their swift, erratic course as they dodged away in the general direction of the salt flats. They landed before he could bring his glasses to bear but he reckoned there were seven Japanese snipe and two painted snipe. The seven would not long have arrived from their northern breeding grounds in Japan but the two Australian painted snipe were probably the same pair that came to nest each season somewhere within the freshwater swamp or on one of the several patches of bare ground around its margins.

The higher land between the boy and the sea had now merged into the zero level of the tidal flats, the breeze flowing in from the east bringing the tang of salt and the strong scent of tough, reedy tussocks in the increasing heat of the sun. The whistling eagles, patrolling the estuary high up, had been calling regularly and the boy spotted a swamp harrier flying low over the water. He decided to skirt the estuary until he came to the dunes, then climb the dunes and return to the house along the beach.

A wood-teal flew past him, almost certainly the one he had seen on the islet and now travelling along a line ruled from there to the mouth of the estuary. Swinging up his glasses he watched the flyer become steadily smaller, its sturdy, somewhat stubby wings whirring. It went as straight as a bullet and at a speed well above that of other wild ducks, though it was in fact one of the smallest of them. On reaching the estuary it planed swiftly down and disappeared below the bank.

Keeping his glasses on the same place the boy took the chance to improve their focus. It was usual—a matter of routine, almost—for the wood-teal to change guard in the early morning and the watcher, aware that most bush creatures are creatures of habit, waited.

Sure enough, before long there came a movement from beyond the she-oaks—a signal quickly lost, then picked up again as the other flyer reappeared, to be caught, head on, with drumming wings in the centre of the boy's field of vision. Steady hands held the subject fully in view —a perfect close-up of projectile-like flight.

The wood-teal continued to come on directly towards where he stood, its image growing larger with every wing-stroke. Then something happened, something inexplicable, as the bird violently changed course, aiming for the nearest point of the creek. A moment and it had disappeared.

He swept the salt flats and the sky and the low shores of the estuary for an explanation but could find none and was about to lower his glasses when he again caught sight of the flock of snipe. They were flying towards him as though returning to the place from where he had first flushed them. Then they, too, exactly as the wood-teal, suddenly flew faster and even more erratically than before. Because the snipe were coming too close and travelling too fast to be held properly in focus the boy decided to lower the glasses, when he noticed the shimmer of other wings in the background.

The newcomer was larger than the snipe, about the same size as the

143

wood-teal, but of very different build and general appearance. Its wings were long and narrow and it flew with such whipping, surging power that its quarry was overtaken with consummate ease. The boy dropped the glasses from his eyes as he heard the thud of a blow and the chance timing of his action was so perfect that the last picture through the lenses was of a burst of feathers frozen against the blue sky and the victim whirling helplessly in mid-air. Then, incredulously, he saw the body of the dead bird bouncing and slithering through the grass almost to his feet. An upraised wing quivered and sank slowly down; a puff of feathers drifted away on the breeze.

Too astounded to feel any immediate sympathy for the loser of that dash across the open flats, the boy switched his attention to the long-winged killer—a stranger to those parts, he felt certain—which, having swept around in a tight curve, was now settling beside its victim. Boldly it came to earth and, when the boy took a step towards it, stood its ground, its fierce stare both unwinking and unafraid.

The boy paused, deeply interested and enthralled at the nearness of this alien, the like of which he had never seen before. A hawk, clearly—a hawk of some kind. It was rather larger than a magpie and even more powerfully knit. The claws were heavy and armed with sharp, curved talons, the legs were very strong and set wide apart, like a bulldog's.

'Get,' the boy muttered, but half-heartedly, for he was still more intrigued by the audacity of the intruder than he was indignant at the violent death of an acquaintance.

'Go on, get out,' he repeated, and swung the glasses by their strap.

The hawk arched its wings angrily and lowered its head. There seemed to be no fear in the creature, only defiance.

Again diverted by his absorbing interest in every inhabitant of the bush the boy was delving deep into his memory. He had seen an illustration, in colour, somewhere; the dark, smoke-grey back and wings, the white throat, the black head and cheeks, as though it were wearing a mask, the cream-coloured breast merging into bars of soft brown, the strong, hooked beak and, most of all, the impression of power given by the wide-set legs, as heavily muscled as a gamecock's.

'A falcon,' the boy whispered, remembering. 'A peregrine falcon.' Then, as though the harshness of this killing had at last dawned upon him, he advanced upon the marauder, threateningly swinging his field glasses. The falcon beat into the air, but not promptly. It was chattering loudly. Because of the speed and ferocity of its attack on the painted snipe and because of its present hostility, the boy was aware also of a certain sense of danger; but he kept on. Again he swung the heavy field glasses, this time vigorously around his head, and, as they hissed through the air, the falcon rose higher, effortlessly, then yet higher. It had ceased to protest, but merely hovered, watching.

The boy picked up the dead bird and examined it, the scientist once more pushing the emotional boy into the background. The body lay warm in his hand, the head dangling. The crown of the head was dark brown with a buff line; and the back also was brown, beautifully mottled and dotted with black. He turned the snipe over. Throat and chest were of the same dark brown, shading to a white abdomen, and the wings were brown, spotted with white and buff and black. As a specimen the falcon's quarry was without blemish. Even the long, slender bill, so ideally suited for probing the soft mud of the marshy flats, was without mark or imperfection in spite of the relative violence of the flyer's crash to earth. Plump, yet trim and streamlined; firm, but with the soft feel of down under its sheath of resilient outer feathers, the snipe seemed to be ideally fashioned for speed of flight—even though it had just been fatally outmatched. Subdued in colouring, the richness of its plumage and the delicacy of its markings could be fully appreciated only at short range.

He weighed the body experimentally, regretfully; then leapt back. But he was too late! A stiff wing-feather whipped across his cheek, a single point of the falcon's claws stung his hand and the painted snipe had gone. The boy shouted. It was a cry of indignation, of admiration —even triumph—for the sheer audacity of the act was to him almost unbelievable. He was relieved, too, for in that moment of shock with the predator flying in his face it had flashed into his mind that his eyes might be the target. Now, however, the falcon was beating its way across the sunny paddock, its kill trailing from its talons.

He groped for the glasses hung around his neck and brought them to bear precisely, so that the bird of prey, already almost out of sight, came distinctly into view again. It was still flying fast, but lower, as though its prize were growing heavier. Its course was on line for the patch of rainforest below the highest, and nearest, point of the escarpment. On approaching the barrier of trees it started to climb laboriously, then veered away and skirted the tall timber, flying on until it was no more than a rhythmic flicker of movement in the glasses.

The boy scrutinised the prow of the escarpment frowning down upon the rainforest and overlooking the quiet, green pastures between

145

the rainforest and the sea. The falcon must have a nest somewhere on the line of its flight, otherwise it would not have carried its heavy burden so far—its nest might be in one of the gullies of the rough country behind the escarpment, or even be on the cliff face itself.

There was blood on his right hand. It had trickled from the base of the thumb across his wrist. The cut was short but quite deep, as clean as though made with a scalpel. He grinned suddenly as he sucked it, wondering how many microscopic portions of how many birds and small mammals he was extracting simultaneously.

The sun was hot as he climbed the ridge again and continued on his way towards the estuary, where he turned left, plodding up the dunes to where the nor'-easter blew cold on his perspiring skin. There were gulls on the beach and terns on the outer rocks of a small reef. Gannets were fishing beyond the breakers—too far out to be watched in comfort, though the boy put the glasses on them and waited for three or four of their spectacular crash dives. The crows had gone. They seldom returned to the beach after their usual early morning forage. High above him a white-breasted sea eagle drifted on motionless wings, a grey-blue V in the sky like some pale-coloured butterfly always about to alight, always in fact in the last buoyant pause before settling—and always wafting lightly on.

'Another of those confounded chicken-hawks', his father would have said. 'I'll shoot it one day, when the rifle's handy', and the boy could never quite decide whether his father still hoped one day to carry out his threat or whether he had accepted the explanation that sea-eagles lived mostly on dead fish washed up by the sea or stranded in the estuaries. After all, as the boy had often, and earnestly, pointed out, how many chickens would an eagle expect to find on a beach? Isn't that so? Ever heard of a sea-eagle hunting inland? He had not argued with his father for many months. He hoped they would not again soon be at loggerheads over the question of his interest in the native fauna.

His mental re-enactment of the last discussion—argument, perhaps —was cut short by a harsh cacophony from the gulls and terns, now somewhat behind him, and he swung round quickly to see what had alarmed them. The air seemed to be as full of white wings flashing as it was full of sound. Then he caught sight of the falcon. It was coming in low and fast over the dunes, its long narrow wings thrusting it along at a speed which, though great, had something of leisureliness about it, too, for the strokes were not finished but clipped short, though vib-

146

rant with nervous strength. Suddenly, the boy gasped and shielded his face. The falcon was flying straight at him. But it curved away low over the sand, then hurtled forward to flash towards the surf, swoop, and snatch one of a group of dotterel almost before the terrified birds had had time to spread their wings.

Gaining height at once the falcon disappeared and the boy ran to the top of a steep, loose-packed dune. As before the falcon was making in the direction of the rainforest but flying normally, quite unencumbered by its diminutive victim.

Two kills in a quarter of an hour! The raider—if both attacks had been made by the same bird—would be a scourge, indeed, to all the smaller birds of the creek, the estuary, the shore and the grassy paddocks. If there were two falcons, and almost certainly there would be, then the situation would be twice as bad.

Near the very limits of sight two other fliers seemed to be harassing the falcon. All three were black against the sky. The boy thought at first that the challengers were magpies then heard, faintly, the ringing cries of the spur-winged plover. The falcon ascended. Fast and daring in their swoops and dives the plovers would soon be hopelessly outdistanced on a rising course by the surging strokes of the falcon. Soon, two of the black specks fell away; the third remained, growing smaller.

When the boy reached home he told his parents, dramatically, of what had happened.

'A damn chicken-hawk,' his father said, exactly as expected. 'They ought to be thinned out.'

'I'll shoot this one myself, Dad. And it's not a chicken-hawk; it's a falcon.'

The man looked astonished. It was the first time his son had ever agreed to exterminate anything.

'Glad you're getting a bit of sense at last. There's that big cove, too. White and grey. Saw him again this morning, over the beach. Leaving out the wedge-tails, he'd be about the biggest chicken-hawk I've seen around these parts.'

The boy was too disgusted to reply. Chicken-hawks, all! From the dainty little nankeen kestrel, living mainly on large insects and never taking anything bigger than a fieldmouse, to the majestic wedge-tailed eagle they were all chicken-hawks to his father. Little chicken-hawks, medium-sized chicken-hawks, big chicken-hawks—chicken-hawks with a wing span of up to ten feet.

'I'll go over early in the morning,' the boy said. 'I think they've got

147

a nest over near the rainforest; maybe on the escarpment.'

'I might come with you. I've got to burn off that lantana thicket I poisoned awhile ago. It's just about right for the fire stick.'

There was only one easy way across the creek to the foot of the escarpment and that was by a path through the lantana and then across a dead tree felled when all the coast was rainforest and which still stretched, solid and undecayed, from bank to bank.

Not that there was ever any pressing reason to get to the other side for there was nothing there except piled-up rocks and the useless tangle of rainforest.

Next day the man did not accompany the boy, after all. There was a fence to fix. So the boy went alone, carrying besides his binoculars a .22 automatic rifle. The poisoned lantana had been reduced to a tangle of dead sticks—an interlaced pile of dead stuff sixty paces long, forty paces through and about twice the height of a man. Snapping off brittle twigs as he went he pushed into the thicket. Probably, no one had used the narrow path since his last school holidays and he went carefully, especially in crossing by the dead tree for the drop down to the creekbed was considerable.

On the other side was a narrow, rock-strewn flat with, beyond that, the slopes of detritus; little stones to slip and roll under the feet and great, jagged chunks embedded in rubble. He had intended to keep close to the cliff's base but found the going too difficult so he went back to the edge of the creek and soon into the sombre shadows of the rainforest. Certainly that way was no less rough, but at least it was stable, the chaos of rock having settled into a solid conglomorate mass; also, he was able to haul himself over many an obstacle by gripping a vine or a branch and, in this manner, scrambled steadily on.

He was resting, out of breath, when he heard the familiar chattering call of the falcon. But the call was different, almost in undertones.

Immediately upon emerging from the trees he spotted the nest— or, rather, the ledge where the nest was; for he had seen a wing tip waving and the chattering had grown in volume, as though an adult bird were feeding young. A pair of falcons and some nestlings would be his guess.

Coming further into the open the boy rested his back against a great block of stone the better to scan the cliff face, bright in the morning sun. He raised his glasses and discovered that one of the adult falcons was staring down at him, its eyes intent in its strange-looking, masked

148

head. But it seemed neither to be alarmed by, nor greatly interested in the boy's scrutiny, and soon settled down to enjoy the sun's warmth.

It was an easy mark. The boy laid down the glasses and picked up the rifle. The target had moved slightly but was still unmissable. He took aim. Or perhaps it would be better to kill the young ones first! If he shot the adult birds the young ones might well survive if they were old enough to fly—if they were not old enough they would starve. Naturally. He lowered the rifle, because it depended on how old the young ones were. He would have to decide the point before shooting.

The other adult falcon—the one the boy could not see because it was sitting towards the back of the ledge—suddenly lifted into the air, fluttered a while, then set off swiftly and vanished behind the cliff's shoulder. That was interesting. It must be hunting the back country, in preference, apparently, to the pastureland.

They waited in silence—the boy in the shadow of the rainforest, the falcon in the sun, drowsing, hunched-up, the nestlings, or nestling, not in evidence, either by sight or sound. It was not long before the hunting falcon was back. The boy thought it was the female. She had killed a quail. The young falcons, two of them, came alive noisily, vehemently, voraciously. A reddish-brown patch below the victim's breast, indicated it was not a stubble quail, but a smaller quail from the high heathland behind the escarpment. The boy had been up there a few times; and a flat, cheerless place it was, swampy in parts and covered with low, coarse grass and dense, tough bushes about two feet in height. There were always king quail there, he knew, at this time of the year.

To judge by the two wide-beaked, featherless heads that had appeared for a moment over the rim of the ledge the young falcons were still some way short of flying. So it would be good tactics, to shoot the parents at once and leave the fledglings to starve. Thus the whole family would be wiped out. Yet already even the most harmless hawks were rare in the district. Being generally unafraid of man they were easy marks for anyone with a gun and it was seldom these days that even a kestrel, formerly so numerous, or a kite was found in that part of the coast. Sea-eagles, harriers and ospreys were rarer still and the wedge-tailed eagle sailed overhead not more than four or five times every year. These were the only falcons he had ever seen and he was beginning to realise how loath he was to carry out the execution.

But he picked up the rifle again. The male falcon, still basking in the sun, blinked at him as he moved. Then the female left on yet

149

another mission into the hinterland. She would fly fast, he would imagine, at about tree-top level, sometimes threading through the high branches, or perching at some vantage point, ready to pick off the unwary traveller in one paralysing burst of speed. Broken-up terrain would suit such a huntress better than open country where the prospective quarry could see danger when it was still in the distance.

She came back with a parrot, partly devoured; whereupon the male falcon made his first excursion for some time, also flying inland. He was away for about fifteen minutes and came back with nothing. Both birds then left together, the male rising up and over the cliff's summit. They were away for longer on this occasion, the male returning first but again with nothing though he had fed, for there were feathers adhering to his talons.

Taking up his rifle and field glasses the boy made his way back to the house.

'Finish off the chicken-hawks?' his father asked expectantly.

'No. I—er—I want to watch 'em a while. They seem to do most of their hunting in the back country.'

'Huh,' the man muttered. 'How's the hand where your mate clawed you?'

The boy flushed but remained silent.

Next day after lunch they set out to burn off the lantana.

'Bring the rifle,' the man directed, 'I'll clean out that hawk's nest myself, before we set a match to that lantana.'

They walked down without speaking, under the highriding sun. There were no birds in the sky, or calling from the trees, or from the ground. The falcons would probably be taking a siesta, for their ledge far up the cliff would be in shade by now, the top of the escarpment being slightly overhanging. Heat waves danced in the distance; an insect skirred loudly, and finished with a metallic click, click, click, over and over again. It was hotter still as they followed the airless path between the high walls of dead lantana. They reached the tree lying across the creek.

'Give me the rifle,' the man said. 'And don't fire the lantana till I get back. I don't want to spend the night over that side of the creek, waiting for the ground to cool.'

The boy handed over the rifle.

'Coming?'

'No.'

150

So the man crossed the dead tree bridge alone and began the long, difficult walk along the base of the cliff.

'Hey!'

The man turned inquiringly.

'Don't forget to load,' the boy finished.

The figure on the other side of the creek did not stir for a while, then came slowly back, stopping at the other end of the dead tree.

'So the magazine's empty?'

'You just said to bring the rifle, Dad. And you've always told me never to load a rifle till you're about ready to use it. Besides, I thought...' His voice trailed off.

His father nodded. He nodded again, several times, as he re-crossed the tree trunk.

'If it wasn't such a hot afternoon I'd send you back for 'em. As it is I'll do the whole job myself.'

Leaning the empty rifle against a nearby tree the man paused to mop the perspiration from his face and throat. 'Can't be expected to put up with pests like that around the place. After all, they're birds of prey! On the look out for what they can pick up—chickens and the like.'

'We haven't got any chickens,' the boy shouted, then stopped.

The man glanced down at him, made as though to speak, but said nothing.

Relieved, the boy changed the subject by saying that it would be safe to set alight to the lantana as the fire could not possibly spread, the grass being so lush. The man thanked him gravely for his advice and lit the western end of the patch so that the flames would burn back into the nor-easter. The growth had been less exuberant at that end, but the flames were soon crackling loudly. A familiar shape, swift and vibrant, flashed through the smoke and swung high into the air, chattering shrilly. Then another as dynamic as the first.

'It's the smoke,' the boy explained quickly. He quoted, 'Hawks are always attracted by smoke because fire drives out the ground birds and animals.'

The man was watching the falcons keenly and, as far as the boy could make out, without hostility. One of the winged hunters plunged into long grass on the margin of the dead thicket and emerged with a native rat. It flew back towards its nest on a ledge of the escarpment.

The fire had now enveloped all the lantana and was burning furiously, a swirl of red-hot fragments spinning up with the smoke.

Driven back by the sudden heat the man moved away, instructing the boy to follow. They continued on for some distance up the rise. The bronzewing pigeons were well out of range of the fire and so it was probably the sound of footsteps coming nearer that put them up. At any rate they rose with a whirr of wings to skim uphill at a tremendous pace, making for the shade trees around the house. Both the boy and his father watched them from the moment they first left the ground. And, though they were hovering some distance further away—near the fire—the falcons must have seen them too. Involuntarily the man ducked twice as, with a double rip of cloven air, one marauder then the other flashed close by his head. The first kill was a matter of straight lines meeting, the falcon homing on its terrified quarry with the in-exorable certainty of a guided missile. A burst of feathers, then a dead bird sousing into the long grass. Swift as a swordstroke the huntress swung round and alighted.

But the second bronzewing, in a desperate effort to escape, shot skyward and the second falcon missed its strike. It was that error of judgment, in fact, that gave the pursuer the chance to show off his astonishing flying ability; he towered majestically, then plummeted down so fast that again the actual blow of the hind claw could not be seen. Only the inevitable puff of feathers and the victim spinning to earth. The man admitted slowly, as he watched the falcons flying off with their booty, that he had certainly never witnessed anything like that before.

'It was the fire that brought 'em, Dad,' the boy explained. 'Usually, they do their hunting in the back country.'

'So you said before.'

The fire had burned down in an hour, leaving piles of white ash. Solid walls of heat prevented close inspection but it was soon decided there was no danger of other fires being sparked off and the walk back to the house was begun.

'Be black out by morning. A man'll be able to get through it without any trouble. And I won't be forgetting the cartridges. Judging from the way those chicken-hawks mowed down the bronzewings I'd say they'd be able to kill anything around this farm, except maybe the full-grown cattle.'

The boy's mouth set. He had been on the point of asking that the extermination of the falcons be postponed indefinitely in order to find out whether they did actually do most of their hunting in the back country. But he decided to say nothing. As for the man, he was fully

aware that further discouragement of the boy's profitless interest in birds might fast lead to the old estrangement between them. But he would concede nothing.

Later that night, about eight, it started to blow.

'I suppose the fire is all right,' the man murmured uneasily. 'Nothing much to burn. I think I'll stroll down some of the way and have a look, all the same.' He unhooked his spectacles, put down his newspaper.

'I'll go,' the boy offered.

His father eased back into his chair again, and nodded his approving thanks.

On his way past the outhouses the boy threw a long handled shovel over his shoulder.

The track to the dead tree crossing was faint because it was not often used but the night was clear and the moon was bright. White ash was banked like snowdrifts. When the wind whirled the flakes from the bigger drifts a deep red glow would appear, then fade again into the general whiteness. Heat seared the boy's face as he came closer but he found a way through the ash at the place where there had once been a path and reached the end of the dead tree bridge. The weathered butt of it was scorched but not burned.

With the long-handled shovel held in front of him he ran at the nearest of the mounds of ash and plunged the shovel into its base, to reel back under the blast of heat. Very soon, however, he was piling the red hot coals on the dried and weathered timber. Almost at once the smoke started to rise and, occasionally, small blue flames would flicker up. Ten minutes later, as he brushed the coals aside, a rain of red sparks fell into the creek and went out. He scooped ash and charred wood out of the tree trunk, using the point of the shovel. Then he went back for a second supply of coals which he poured into the hollow he had made. Fanned by the wind the tree started to burn freely. There was not much doubt that it would quickly burn through, perhaps before midnight.

Returning up the long, gradual rise at his best pace he put the shovel back in the shed and re-entered the house. The fire, he said, was safe.

His father left next morning to exterminate the falcons and their brood; but he left rather late. Nor did he go right to the bank of the creek but stopped at the edge of the devastated area. Then he came back up the hill, cleared the magazine of the .22 automatic and put it away.

'A spark must have lodged on the dead tree,' he remarked matter-of-factly. 'It's down in the creek, burned through.'

The boy wondered then if the man were glad of an excuse to let the falcons live—at least until he knew more about them. Perhaps, he, too, had been fascinated by the dash and fearlessness of the birds of prey. That might have been the reason why he had asked no questions about the destruction of the dead tree bridge.

Then he remembered that his father had used the long-handled shovel that very morning, and must surely have noticed then that its blade was streaked with grey ash.

154

*Some other*
REED BOOKS
*are described overleaf*

A PORTFOLIO OF AUSTRALIAN BIRDS

by William Cooper and Keith Hindwood

Although William Cooper's bird illustrations have already gained for him an enviable reputation in his field, this is the first time that his exquisite paintings have been made available in book form. In twenty-five superb full colour plates he has captured, with an imaginative precision that no photograph could possibly emulate, the special character and background of a wide variety of Australia's more interesting and colourful birds.

Keith Hindwood, an experienced naturalist and the author of *Australian Birds in Colour,* has written a lively, comprehensive and informative text about each of the birds illustrated and about Australian bird life in general.

Scientifically accurate, with a meticulous attention to detail, the paintings clearly protray the various birds in their natural surroundings. Among the species illustrated and discussed are the Golden Whistler, the Eastern Whipbird, the Rainbow Lorikeet, the Regent Honeyeater, the Red-tailed Black Cockatoo, and a number of other highly interesting birds.

This is more than just another bird book. The illustrations stand on their own, and will appeal to art lovers as well as to those who wish to learn more about some of Australia's most beautiful birds.

14½" x 11⅛", 64 pages, 25 full-colour art-plates, cased, jacket.
$9.95

## NATURE WALKABOUT
### by Vincent Serventy

A family caravan tour from Perth to Darwin and Alice, up to Australia's north-eastern point, south via Brisbane and Sydney. The journey took six months to complete and covered 15,000 miles. The writing style is simple and conversational in a unique Serventy style, but is full, too, of facts and interesting information about Australia's incredible wildlife. This is a journey every armchair traveller will want to make, and will be able to see as well, in the dozens of beautiful colour plates.

9½" x 6", 140 pages, 32 pages of full-colour photographs, cased, jacket. $5.50

## SOUTHERN WALKABOUT
### by Vincent Serventy

This title completes the circuit of Australia by the Serventy family, commenced in *Nature Walkabout* above. This trip starts at Sydney, travels through Victoria, South Australia, and across the Nullarbor to Perth, with many side-trips to fascinating islands such as Rottnest and the Abrolhos. Of prime appeal to all nature lovers, teachers, and zoologists, this book will also entertain many an armchair traveller. With the usual glorious array of colour photographs.

9½" x 6", 160 pages, 32 pages of full-colour photographs, cased, jacket. $5.50

## QUINKAN COUNTRY
### by P. J. Trezise

Adventures in search of Aboriginal cave paintings. A rare combination—a serious artistic and anthropological study which is also a spanking good story. The discovery by the author of beautiful and imaginative cave paintings in the remote interior of Cape York Peninsula prompted Dick, Harry, Caesar, and other Aboriginals to explain to him the tribal legends on which the paintings were based. Woven into an anthropological study are these delightful stories, together with the adventure surrounding the various discoveries.

9½" x 6", 164 pages, 32 pages of full-colour photographs, cased, jacket. $5.50